Applications of
Molecular
Biology
in
Environmental
Chemistry

Edited by
Roger A. Minear
Allan M. Ford
Lawrence L. Needham
Nathan J. Karch

LEWIS PUBLISHERS
Boca Raton   New York   London   Tokyo

Library of Congress Cataloging-in-Publication Data

Applications of molecular biology in environmental chemistry / edited
    by Rober A. Minear ... [et al.]
            p.   cm.
    Includes bibliographical references and index.
    ISBN 0-87371-951-4
        1. Genetic toxicology.   2.Molecular toxicology.   3. Environmental
chemistry.   4. biological monitoring.   5. Environmental health.
    I. Minear, R. A.
    RA1224.3.A67   1995
    615.9¢02--dc20   94-38089
                                                                     CIP

© 1995 by CRC Press, Inc.
Lewis Publishers is an imprint of CRC Press

No claim to original U.S. Government works
International Standard Book Number 0-87371-951-4
Library of Congress Card Number 94-38089
Printed in the United States of America   2 3 4 5 6 7 8 9 0
Printed on acid-free paper

During the 20th century, we have experienced a shift in the leading causes of death from infectious diseases, such as pneumonia and influenza, to chronic diseases, such as heart disease and cancer. This decrease in deaths from infectious diseases is due in large part to improved public health measures such as vaccines (to prevent the onset) and antibiotics and other drugs (to treat the disease). However, whereas infectious diseases are often related to a single infectious agent, chronic diseases are often related to a combination of environmental (including occupational) chemical exposures and genetic factors. Analytical chemists are at the forefront in identifying these chemical pollutants and tracking their transport and fate throughout the environment. They also measure the concentration of these pollutants in environmental matrices such as air, water, food, and soil. Once the pollutant enters the human body, the analytical chemist measures the pollutant, its metabolite(s), or additional products. This internal dose or biologically effective dose is a biomarker of exposure. However, the analytical chemist is but one part of a team. Other team members, including epidemiologists, toxicologists, statisticians, molecular biologists, and clinicians, study the relationship between exposure, biological responses, and clinical outcomes; in so doing, they look for biomarkers of effect. As part of this exercise, which is designed to provide understanding of the biological mechanisms involved in diseases possibly induced by environmental pollutants, they also examine measures of susceptibility markers, which characterize an individual's or a population's inherent sensitivity to the pollutant. Much work has been accomplished in these areas.

However, much more work is to be done. Analytical chemists need to develop and apply improved analytical methods to measure the pollutant, its metabolite(s), and its various possible adducts in humans. We need better measures of the amount and activity of the pollutant in environmental specimens. We need better subclinical measures of effect and improved methods to identify and measure susceptibility markers in large populations. This book is derived from two American Chemical Society Symposia: (1) Human Biomonitoring, August 1991, and (2) Molecular Biological Tools in Environmental Chemistry, Biology and Engineering, August 1992. Some of the latest work as of this time designed to assess potential exposure (i.e., environmental concentration and activity of pollutants), dose to humans, and a molecular basis for some of the effected biological mechanisms, is presented. This field is a rapidly changing one, yet many new developments build upon existing state-of-the-art. It is our intent to convey a picture in the time continuum. Some, but not all, chapters reflect new information available after symposium presentations.

The Editors

Roger A. Minear, Director of the Institute for Environmental Studies and Professor of Civil Engineering at the University of Illinois, received his B.S. (1964) in Chemistry, M.S.E. (1965) in Sanitary Engineering, and Ph.D. (1971) in Civil Engineering (specializing in environmental chemistry) from the University of Washington. He served as an Instructor at Oregon State University, 1966–1967, an Assistant Professor at Illinois Institute of Technology, 1970–1973, joined the University of Tennessee, Knoxville, as an Associate Professor in 1973, became Professor in 1977, and was named the first Armour T. Granger Professor in 1983.

Dr. Minear holds membership in the American Chemical Society, the American Society of Civil Engineers, the American Society of Limnology and Oceanography, the American Association for the Advancement of Science, the American Water Works Association, the Association of Environmental Engineering Professors, the International Association on Water Quality, the Society of Environmental Toxicology and Chemistry, the Water Environment Federation, and Sigma Xi.

Dr. Minear's major areas of scientific interest involve the nature, origin, transport, and transformation of organic and inorganic compounds in natural and wastewaters; chemistry of aqueous solutions and chemical processes of water and wastewater treatment; and trace and environmental analysis. He has generated 101 professional paper presentations, 42 invited seminar presentations, 103 published articles, chapters, reports, and 33 graduate theses in the Environmental Science and Engineering field.

Allan M. Ford is currently Director of the Gulf Coast Hazardous Substance Research Center, a consortium of eight Gulf Coast universities with research programs into the remediation of hazardous waste sites, pollution prevention, and environmental sociology. Dr. Ford was Director of Monsanto Company's Environmental Sciences Center from 1986 through 1992. He received his Bachelor's Degree in Chemistry from Iowa State University and his Ph.D. in Analytical Chemistry from Kansas State University.

Dr. Ford has served in numerous capacities within the St. Louis Section of the American Chemical Society including chairman and councilor. On a national level, he is an appointed member of the Committee on Environmental Improvement and chairs their Task Force on Environmental Research Funding. He has been an elected member at large of the American Chemical Society's Environmental Division. He is also a member of AAAS, SETAC, AICHE, and the advisory board of the Committee for the National Institute for the Environment.

Larry L. Needham is Chief of the Analytical Toxicology Branch in the National Center for Environmental Health of the U.S. Centers for Disease Control and Prevention in Atlanta. His laboratory is heavily involved in the development and application of analytical methods for assessing human exposure to environmental toxicants — including polychlorinated dibenzo-p-dioxins, polychlorinated biphenyls, persistent and nonpersistent pesticides, polyaromatic hydrocarbons, and volatile organic compounds. His laboratory has performed laboratory measurements in numerous epidemiologic investigations, including Vietnam veterans, occupationally exposed workers, residents around Superfund sites, general population, and residents along the U.S. and Mexico border.

Previously, Dr. Needham was an Assistant Professor in the School of Pharmacy of Auburn University and was a Research Chemist with the General Electric Company. He received his Bachelor's Degree in chemistry from Middle Tennessee State University and his Ph.D. in organic chemistry from the University of Georgia. He received postdoctoral training both at the University of Georgia and Vanderbilt University. Dr. Needham has authored or co-authored more than 125 published articles.

Nathan J. Karch is the president of Karch & Associates, Inc., a consulting firm that specializes in toxicology, epidemiology, and risk assessment. For nearly 20 years, he has managed a variety of complex projects, including assistance to private clients or their attorneys in evaluating scientific issues raised in product liability, toxic tort, and related litigation, in reviewing the toxicity of various industrial chemicals and mineral products, in preparing comments on various regulatory proposals, and in evaluating worker risks and industrial hygiene practices.

Dr. Karch has directed projects at Karch & Associates over the past 10 years in which the exposure, fate, process chemistry, health effects, and risks associated with chemicals were evaluated, including over 30 very large lawsuits involving multiple plaintiffs. He has evaluated and coordinated assessments of medical, toxicological, epidemiological, and environmental issues, as well as physical/chemical properties and other modeling information. Some of the chemicals at issue are chlorinated dibenzodioxins (e.g., TCDD) and dibenzofurans, pentachlorophenol, polychlorinated biphenyls (PCBs), polycyclic aromatic hydrocarbons (PAHs), lead, cadmium, chromium, barium, 2,4-D, 2,4,5-T, atrazine, various other herbicides and pesticide chemicals, tetrachloroethylene, trichloroethylene, ethylene oxide, formaldehyde, and hydrogen chloride. For the government, Dr. Karch has directed contracts for the Occupational Safety and Health Administration and the Environmental Protection Agency.

Dr. Karch directed a large project for rail clients that involved the development of a unique model for ranking the potential for acute lethality and long-term adverse health effects of chemicals in over 750 hazardous materials carried by rail nationwide. The project focused upon inhalation hazards to communities adjacent to a large spill of a volatile chemical. An aspect of this project involved presentation of the methods of evaluation and relevant conclusions to the U.S. Environmental Protection Agency and the U.S. Department of Transportation to assure consistency between the agencies in their approach to inhalation hazards.

Dr. Karch has presented scientific papers on a range of topics involving the interpretation of toxicological and epidemiological data, regulatory issues, and risk assessment. Before starting Karch & Associates, Inc., Dr. Karch was a Senior Science Advisor at Clement Associates. He has served as acting senior staff member for toxic substances and environmental health at the Council on Environmental Quality in Executive Office of the President during the Carter Administration. He was a senior staff officer at the National Academy of Sciences before entering government, and a legislative assistant at the American Chemical Society after receiving his Ph.D. in physical organic chemistry from Yale University in 1973. Dr. Karch took postgraduate courses in toxicology, epidemiology, and biostatistics, and he is certified by the American Board of Toxicology.

Contributors

David L. Ashley
Toxicology Branch
Division of Environmental Health
  Laboratory Sciences
National Center for Environmental
  Health
Centers for Disease Control and
  Prevention
Atlanta, Georgia

Ronald M. Atlas
Department of Biology
University of Louisville
Louisville, Kentucky

Charles E. Becker
University of California-San
  Francisco
Center for Occupational and
  Environmental Health
San Francisco, California

Asim K. Bej
Department of Biology
University of Alabama
Birmingham, Alabama

Robert A. Bethem
Alta Analytical Laboratory, Inc.
El Dorado Hills, California

Michael A. Bonin
Toxicology Branch
Division of Environmental Health
  Laboratory Sciences
National Center for Environmental
  Health
Centers for Disease Control
Atlanta, Georgia

F. Reber Brown
California Public Health Foundation
Hazardous Materials Laboratory
Berkeley, California

Frank O. Bryant
United States Department of
  Agriculture
Agricultural Research Center
Athens, Georgia

Frederick L. Cardinali
Toxicology Branch
Division of Environmental Health
  Laboratory Sciences
National Center for Environmental
  Health
Centers for Disease Control
Atlanta, Georgia

Thomas A. Cebula
Division of Molecular Biological
  Research and Evaluation
Center for Food Safety and Applied
  Nutrition
Food and Drug Administration
Washington, D.C.

Albert M. Cheh
Department of Chemistry
American University
Washington, D.C.

Ruey-Hwa Chen
Carcinogenesis Laboratory
Department of Microbiology and
  Department of Biochemistry
Michigan State University
East Lansing, Michigan

David Cortez
Institute for Environmental Studies
University of Illinois at Urbana-
    Champaign
Urbana, Illinois

Horace G. Cutler
United States Department of
    Agriculture
Agricultural Research Center
Athens, Georgia

P. M. DiGrazia
Center for Environmental Biotechnology
University of Tennessee
Knoxville, Tennessee

Joseph L. DiCesare
Perkin-Elmer Cetus
Norwalk, Conneticut

William M. Draper
California Department of Health Services
Berkeley, California

P. M. Fedorak
Department of Microbiology
University of Alberta
Edmonton, Alberta, Canada

Dennis K. Flaherty
Monsanto Company
The Agricultural Group
Environmental Health Laboratory
St. Louis, Missouri

James T. Fleming
Center for Environmental Biotechnology
Knoxville, Tennessee

J. Foght
Department of Microbiology
University of Alberta
Edmonton, Alberta, Canada

Frank J. Gonzalez
Laboratory of Molecular
    Carcinogenesis
National Cancer Institute
National Institutes of Health
Bethesda, Maryland

J. G. Guillot
Centre de Toxicologie du Quebec
Ste-Foy, Quebec, Canada

Lawrence A. Haff
Perkin-Elmer Cetus
Norwalk, Connecticut

A. Heitzer
Center for Environmental Biotechnology
University of Tennessee
Knoxville, Tennessee

T. R. Jack
Novacor Research and Technology
    Corporation
Calgary, Alberta, Canada

Donald M. Jerina
Laboratory of Bioorganic
    Chemistry
National Institute of Diabetes
Digestive and Kidney Diseases
National Institutes of Health
Bethesda, Maryland

A. LeBlanc
Centre de Toxicologie du Quebec
Ste-Foy, Quebec, Canada

Chuang Lu
Department of Chemistry
American University
Washington, D.C.

Veronica M. Maher
Carcinogenesis Laboratory
Department of Microbiology and
    Department of Biochemistry
Michigan State University
East Lansing, Michigan

Shawn McCarty
Louisville Water Company
Louisville, Kentucky

J. Justin McCormick
Carcinogenesis Laboratory
Department of Microbiology and
    Department of Biochemistry
Michigan State University
East Lansing, Michigan

Joan M. McCraw
Toxicology Branch
Division of Environmental Health
    Laboratory Sciences
National Center for Environmental
    Health
Centers for Disease Control
Atlanta, Georgia

Michael J. Miille
Quanterra Enviromental Services
Englewood, Colorado

Rashmii Nair
Monsanto Company
St. Louis, Missouri

Larry L. Needham
Toxicology Branch
Division of Environmental Health
    Laboratory Sciences
National Center for Environmental
    Health
Centers for Disease Control
Atlanta, Georiga

Robert Orth
Monsanto Company
St. Louis, Missouri

Michael J. Plewa
Institute for Environmental Studies
University of Illinois at Urbana-
    Champaign
Urbana, Illinois

O. Samuel
Centre de Toxicologie du Quebec
Ste-Foy, Quebec, Canada

Gary S. Sayler
Center for Environmental
    Biotechnology
University of Tennessee
Knoxville, Tennessee

Shannon R. Smith
Institute for Environmental Studies
University of Illinois at Urbana-
    Champaign
Urbana, Illinois

Gary Spies
Monsanto Company
St. Louis, Missouri

A. J. Telang
Division of Biochemistry
Department of Biological Sciences
University of Calgary
Calgary, Alberta, Canada

Margaret J. Timme
Institute for Environmental Studies
University of Illinois at Urbana-
    Champaign
Urbana, Illinois

G. Voordouw
Division of Biochemistry
Department of Biological Sciences
University of Calgary
Calgary, Alberta, Canada

Elizabeth D. Wagner
Institute for Environmental Studies
University of Illinois at Urbana-
    Champaign
Urbana, Illinois

O. F. Webb
Center for Environmental Biotechnology
University of Tennessee
Knoxville, Tennessee

J. P. Weber
Centre de Toxicologie du Quebec
Ste-Foy, Quebec, Canada

Jay M. Wendling
Monsanto Company
St. Louis, Missouri

D. W. S. Westlake
Department of Microbiology
University of Alberta
Edmonton, Alberta, Canada

Joe V. Wooten
Toxicology Branch
Division of Environmental Health
    Laboratory Sciences
National Center for Environmental
    Health
Centers for Disease Control
Atlanta, Georgia

Haruhiko Yagi
Laboratory of Bioorganic Chemistry
National Institute of Diabetes
Digestive and Kidney Diseases
National Institutes of Health
Bethesda, Maryland

Contents

Chapter 6
Molecular Analysis of Frameshift Mutations Induced by Plant-
Activated 2-Aminofluorene 69

**Michael J. Plewa, Margaret J. Timme, David Cortez,
Shannon R. Smith, and Elizabeth D. Wagner**

Chapter 7
Identification of Sulfate-Reducing Bacteria by Hydrogenase Gene Probes
and Reverse Sample Genome Probing 81

**G. Voordouw, A. J. Telang, T. R. Jack, J. Foght, P. M. Fedorak, and
D. W. S. Westlake**

SECTION II. BIOMONITORING AND BIOANALYTICAL CHEMISTRY

Chapter 10
Important Considerations in the Ultratrace Measurement of
Volatile Organic Compounds in Blood 135
**David L. Ashley, Michael A. Bonin, Frederick L. Cardinali,
Joan M. McCraw, Joe V. Wooten, and Larry L. Needham**

Chapter 11
Applications of New HPLC/MS Techniques in Human Biomonitoring 147
**William M. Draper, F. Reber Brown, Robert A. Bethem,
Michael J. Miille, and Charles E. Becker**

Chapter 12
Measurement of Reductive Dechlorination of Pentachlorophenol by
Actinomyces viscosus Strain *dechlorini* by GC/MS Techniques 171
Frank O. Bryant and Horace G. Cutler

Chapter 13
A Versatile Bioluminescent Reporter System for Organic Pollutant
Bioavailability and Biodegradation 191
A. Heitzer, O. F. Webb, P. M. DiGrazia, and Gary S. Sayler

Applications of
Molecular
Biology
in
Environmental
Chemistry

Mutational Spectra and Other Microbiological Processes

Monitoring Microbial Pathogens and Indicator Microorganisms in Water Using the Polymerase Chain Reaction and Gene Probe Methods

Ronald M. Atlas, Asim K. Bej, Joseph L. DiCesare, Lawrence A. Haff, and Shawn McCarty

INTRODUCTION

Potable water supplies are routinely monitored for the presence of coliform bacteria as indicators of possible fecal contamination carrying associated enteric pathogens.[1] The margin of safety for possible contamination by fecal matter in drinking water depends upon the development and use of sensitive, specific, and time-efficient methods for the detection of such microorganisms. Public health measures require bacteriological surveillance of environmental water supplies, e.g., monitoring for levels of enteric pathogens that can be transmitted via the fecal–oral route.[2] Federal regulations require the detection level to be 1 indicator bacterial coliform and 1 confirmed specific fecal indicator *Escherichia coli* cell per 100 mL of drinking water.[1,3]

The traditional methods for detecting coliform bacteria rely upon culturing on a medium that selectively permits the growth of Gram-negative bacteria and differentially detects lactose-utilizing bacteria, e.g., using MacConkey's, m-Endo, eosin methylene blue (EMB), or brilliant-green-lactose-bile media.[1] By using these media and an incubation temperature of 35°C, total coliform bacteria, which includes members of the genera *Escherichia, Enterobacter, Citrobacter,* and *Klebsiella*, among others, are enumerated; using an elevated incubation temperature, e.g., 44.5°C, fecal coliform bacteria are enumerated.[2] *Escherichia coli* is used as the critical coliform indicator for fecal contamination because it is specific for the intestinal tracts of warm-blooded animals[1,3] and because it is present in numbers greater than enteric pathogens in water contaminated with human fecal matter.[4] However, the isolation and confirmed characterization of *E. coli* from a mixture of coliforms usually require several

0-87371-951-4/95/$0.00+$.50

days.[5] Several alternate defined substrate tests have been developed for the detection of E. coli. The Colilert® test (Access, New Haven, CT), for example, combines measurement of β-galactosidase activity using o-nitrophenyl-β-galactopyranoside (ONPG) for total coliforms and detection of β-D-glucuronidase activity using a fluorogenic assay with 4-methylunbelliferyl-β-glucuronidide (MUG) that measures β-D-glucuronidase activity of E. coli to indicate the presence of E. coli.[6–8]

As a rapid, specific, and sensitive alternate procedure, we as well as other groups have been developing noncultural, genetically based methods for the environmental detection of coliforms and specific pathogens based upon the recovery of DNA from water, amplification of target nucleotide sequences specifically associated with the coliform bacteria by using the polymerase chain reaction (PCR), and detection of the amplified DNA with gene probes.[9–15] Because PCR has been shown to permit the detection of a single bacterial cell,[11] the approach has been to avoid the problems inherent in the conventional culturing methods and to develop a simple, sensitive detection method by using PCR and gene probes. We have examined the specificity and sensitivity of this detection method using the target gene lacZ for detection of total coliforms, lamB gene for detection of E. coli, Salmonella, and Shigella; and the β-glucuronidase gene (uidA) as targets for PCR amplification for specific detection of E. coli and closely related Shigella.[12] We have been successful in developing a method for the simultaneous rapid detection of total coliforms and E. coli using the genes (lacZ and uidA) that code for the enzymes that form the basis of the Colilert test. We have also developed PCR-based assays for specific waterborne pathogens—Giardia[16,17] and Legionella[18,19]—and other research groups have developed methods based upon PCR for the detection of enteric viruses.[20]

SAMPLE COLLECTION AND PROCESSING

One of the problems for molecular detection methods is how to ensure capture of a single target microbial gene in a 100-mL sample. Most molecular biologists use less than 1 mL volumes containing 10^3 or greater gene copies. Typical recovery procedures involve centrifugation and DNA extraction and purification. The efficiency of recovery from such procedures is not adequate for water quality monitoring. To ensure detection of single target cells, we have used filtration with 0.5-μm pore size filters to collect bacteria from 100-mL water samples. Filters composed of nitrocellulose or cellulose acetate, which are commonly used to concentrate bacterial cells for water quality analyses, were found to interfere with PCR, forcing consideration of alternate filters or means of cell concentration.[13] Tests with a variety of filters demonstrated that certain Teflon® filters could be used for concentration of bacterial cells from 100-mL samples and would not interfere with DNA amplification even when the filter was retained during PCR. By not removing the filter the collection process ensures no loss of target cells during processing. Initially 25-mm-diameter filters were used, but because PCR is run effectively in small volume reactions

(≤100 mL), 13-mm-diameter filters were subsequently used. These filters permit filtration of low turbidity waters within minutes. Besides collecting cells it is necessary to release DNA from the cells to perform PCR. After testing a variety of chemical lysis procedures, we determined that bacterial cells could universally be lysed by repetitive freeze–thaw cycling or by heating to 95°C in the presence of a detergent.

In a typical test procedure, the water sample is dechlorinated by treatment with 0.1% (w/v) sodium thiosulfate. One hundred milliliters of the dechlorinated water is filtered through an ethanol presoaked 13-mm Fluoropore membrane (FHLP, Millipore, 0.5 μm pore size) using a Swinnex® filter holder and a filter manifold (Millipore). The filter is rolled and transferred using sterile forceps to a 0.6-mL GeneAmp® tube (Perkin-Elmer Cetus® Corp., Norwalk, CT) with cell-coated side facing inward. One hundred microliters of 0.1% diethylpyrocarbonate-treated sterile water is added to the tube and vortexed vigorously for 5–10 sec to release the cells from the filter surface to the liquid phase. Five to ten freeze–thaw cycles are performed using an ethanol–dry ice bath and warm water (45–50°C approximately), respectively.[13]

DNA AMPLIFICATION OF TARGET DNA SEQUENCES

PCR amplification is performed using a DNA thermal cycler and GeneAmp kit with native *Taq* DNA polymerase (Perkin-Elmer Cetus). PCR solution is added to the tubes containing the filters. The PCR solution used contained 1 × PCR amplification buffer [10 × buffer contains 50 mM KCl, 100 mM Tris-Cl, pH 8.13, 25 mM MgCl$_2$, and 0.01% (w/v) gelatin], 200 μM each of the dNTPs, 0.5–1.0 μM of each of the primers, and 2.5 units *Taq* DNA polymerase. Total volume for PCR reaction is 150 μL. The template DNA is initially denatured at 96°C for 3 min, then a total of 25–40 PCR cycles are run using a two-temperature PCR cycle with denaturation at 96°C for 1 min, primer annealing, and extension at 50°C for 1 min.

Several target gene sequences are used to detect different indicator bacteria. For example, the 0.264-kb region of *E. coli lacZ* gene is amplified by using 24-mer primers 5′-ATGAAAGCTGGCTACAGGAAGGCC-3′ and 5′GGTTTATGCAGCAACGAGACGTCA-3′ to detect total coliform bacteria.[11] A 309-bp segment of the coding region of the *lamB* gene of *E. coli* is amplified using 24-mer primers 5′-CTGATCGAATGGCTGCCAGGCTCC-3′ and 5′-CAACCAGACGATAGTTATCACGCA-3′ to detect *E. coli, Salmonella,* and *Shigella.*[11] A 153-bp *E. coli uidR* region is amplified by using 22-mer primers 5′-TGTTACGTCCTGTAGAAAGCCC-3′ and 5′-AAAACTGCCTGGCACAG-CAATT-3′[12] or a 147-bp *E. coli uidA* region is amplified using a pair of 20- and 21-mer primers 5′-AAAACGGCAAGAAAAAGCAG-3′ and 5′-ACGCGTGGT-TACAGTCTTGCG-3′ to detect *E. coli* and *Shigella.*[12]

Multiplex PCR amplification using primers for multiple genes and colorimetric detection of the amplified DNA by immobilized capture probe method is performed[10] to detect simultaneously the different target organisms. For duplex amplification equimolar quantities of the *lacZ* and *uidA* sets of primers (0.5 or 1.0 μM of each of the primers) are used. Biotin is incorporated during PCR amplification by adding biotin-11-dUTP (Sigma) and dTTP with a ratio of 1:3, respectively.[14] The oligonucleotide probes containing homopolymer dT-tail are immobilized by photofixation using ultraviolet light on the nylon membrane.[10]

With known test bacteria, addition of equimolar concentrations of multiple primer sets and incorporation of biotin-11-dUTP during PCR amplification of the *E. coli* genomic DNA resulted in an almost equal quantity of all target DNAs by multiplex PCR. Hybridization of the multiplex PCR-amplified DNAs with the immobilized capture probes using Blue-Gene kit (BRL) showed no nonspecific signal.

SPECIFICITY OF COLIFORM AND *E. COLI* DETECTION

As a target for PCR amplification we selected a region of *lacZ* because conventional coliform monitoring is based upon detection of the activity of the gene product (β-galactosidase) of *lacZ* produced by coliform bacteria. We also chose as a target for PCR amplification a region of the *malB* gene that codes for maltose transport protein because this region includes *lamB*, which encodes a surface protein that is recognized specifically by the *E. coli*-specific phage λ. The reason for using *uidR* and *uidA* of the β-glucuronidase gene was based on the fact that this enzyme is produced by all *E. coli* but not by other coliform bacteria. The hypothesis was that PCR amplification of *lacZ* would occur for all (total) coliforms and that PCR amplification of *lamB* and *uidR* would be specific for the fecal coliform *E. coli*. This hypothesis proved to be, for the most part, correct. Amplification of *lacZ* did detect coliform bacteria and did not detect noncoliform bacteria including the Gram-positive lactose-fermenting bacteria. Thus, *lacZ* is a good choice for a genetically based detection system for total coliforms.

Besides detecting *E. coli*, however, the amplification of *lamB* detected *Salmonella* and *Shigella* and the amplification of *uidR* and *uidA* detected *Shigella*. Thus, PCR amplification of *lamB* and *uidR*, as demonstrated here, provides a means of monitoring the indicator bacterial species of fecal contamination (*E. coli*) and also of the enteric bacterial pathogens that cause waterborne disease outbreaks, most importantly *Salmonella* and *Shigella*. A 0.264-kb DNA band was detected in the agarose gel for the *lacZ* gene target for all coliform bacteria; this band was not observed with any of the noncoliform bacteria. A 0.147-kb DNA band in the agarose gel for the *uidA* target was detected for all *E. coli* and *Shigella* strains. No other coliform or noncoliform bacteria tested showed any amplification. Similarly, a 0.346-kb amplified DNA band was detected for *E. coli*, *Salmonella*, and *Shigella*; no other bacterial species showed amplification with these primers. All environmental *E. coli* isolates—including phenotypically MUG-negative isolates—and *Shigella* ioslates showed positive amplified DNA bands for both *lacZ* and *uidA* targets when multiplex PCR amplification was performed. *Escherichia* species other than *E. coli* amplified with *lacZ* but not with *uidA*. All other coliform bacteria amplified only the *lacZ* target. No amplification was detected with other bacteria.

By multiplex PCR amplification of *lacZ, lamB,* and *uidR* sequences and by immobilized capture probe detection, we have developed a system that detects not only the indicator species (*E. coli*), but also the pathogens (*Salmonella* and *Shigella*) whose presence indicates the bacteriological safety of potable water supplies.

SENSITIVITY OF PCR DETECTION

The sensitivity of detection achieved by amplification of *lacZ, lamB, uidR,* and *uidA* coupled with ^{32}P-labeled gene probes, or by colorimetric immobilized capture probe detection assay, was equivalent to 1–10 ag of target DNA, i.e., single genome copy (single cell) detection. Tests with dilutions of viable cells of coliform bacteria indicated that 1–5 cells per 100 mL of water could be detected. By filtering samples and releasing the DNA by freeze–thaw cycling, we were able to detect coliform bacteria with single cell per 100 mL sensitivity. Thus, the sensitivity of the PCR-gene probe method is adequate for water quality monitoring to meet federal regulations.

FIELD TRIALS

To evaluate the specificity, sensitivity, and effectiveness of PCR-gene probe-based detection methods for water quality monitoring, 90 water samples were analyzed by multiplex PCR using *lacZ* and *uidA* genes as targets, defined substrates, and conventional plate count methods.[9,14] Some of the water samples were from the Ohio River (raw water) and some of them were from Louisville Water Company drinking water purification system (finished water). In this study, none of the finished water samples showed coliform positive by any of the plate count, defined substrate, or PCR tests. Statistical analyses indicated that the defined substrate and multiplex PCR methods were equivalent for detection of total coliform bacteria (Table 1.1). Out of 90 samples tested, in only 7 samples did PCR and Colilert®-ONPG tests differ for total coliform detection. Differences between the two tests usually occurred where the concentration of coliforms was so low that repetitive 100 mL samples would likely show variability.

With regard to the specific detection of *E. coli*, the indices of agreement indicated approximately 80% correspondence when both the PCR- and MUG-based tests detected *E. coli* or when both failed to detect *E. coli*. As with the total coliform test comparisons, many of the differences occurred at low concentrations, which may be due to differences between samples analyzed. The PCR

Table 1.1 Comparison of PCR and Colilert Methods for the Detection of Total Coliform Bacteria and *Escherichia coli* Showing Results of Tests of 90 Samples and Statistical Analyses

| | Test results (number of cases) | | | |
| | Total coliform | | E. coli | |
	PCR+	PCR–	PCR+	PCR–
Colilert +	61	3	34	5
Colilert –	4	22	16	35
	Statistical analyses			
Index of agreement	0.92		0.77	
McNemar	0.14		5.8[a]	
Binomial probability	0.5		0.02[a]	

[a] Statistical difference significant at $p > 0.05$.

uidA detection method, however, showed a significantly greater number of differences where PCR was positive and the MUG test was negative; this was indicated by the significance of the McNemar tests and the variance from the expected exact binomial distribution. This result suggests that MUG-negative strains might be undetected using the defined substrate test and show a positive result by PCR. Considering all the comparisons between the MUG-based defined substrate and *uidA* PCR tests for *E. coli*, if 15% of the *E. coli* in the samples were MUG-negative phenotype, there would be a 99% probability according to the exact binomial distribution of obtaining the results we found. Independent tests have indicated that 15% of Ohio River *E. coli* isolates are phenotypically MUG-negative; French researchers have also reported 15% of their *E. coli* isolates are MUG-phenotype but can be detected by PCR-gene probe methods based upon the *uid* gene target.[21]

SUMMARY

The use of PCR and gene probes permits both the specificity and the sensitivity necessary as a basis for a method for monitoring coliforms as indicators of human fecal contamination of water. The advantages of the PCR system include the sensitivity of PCR detection, which can detect single cells in 100 mL water samples; the specificity of PCR-gene probe detection for target microorganisms; the speed from time of sample collection to completion of analysis (which should take less than 6 hr using nonradioactive probes); and the ability simultaneously to detect multiple target bacteria, which can include both a general indicator and a series of specific target pathogens. Multiplex PCR amplification and further development of nonisotopic gene probe detection techniques, such as immobilized capture probe, can permit a rapid and reliable means of assessing the bacteriological safety of waters and should provide an effective alternative methodology to the conventional viable culture methods. Multiplex PCR and gene probe detection of target *lacZ* and *uidA* genes appear to form the basis for the detection of total coliform bacteria and *E. coli*, respectively. The *uidA*-PCR method for *E. coli* detection appears to detect MUG-negative strains that are not detected by the MUG-based defined substrate tests. A greater number of samples from varying water sources will have to be analyzed to establish the efficacy of the PCR method for water quality monitoring before it can be proposed as an alternate to approved and pending plate count and defined substrate methods.

REFERENCES

1. *Standard Methods for the Examination of Water and Wastewater*, 16th ed. APHA, AWWA, and WPCF, Washington, D.C., 1985.
2. *Bacterial Indicators/Health Hazards Associated with Water*. ASTM, Philadelphia (ASTM Publication 635) 1977.
3. Geldreich, E. E. *Bacterial Populations and Indicator Concepts in Feces, Sewage, Stormwater and Solid Wastes*. Ann Arbor Science Publishers, Orlando, FL, 1983.
4. Bonde, G. J. Bacterial indicators of water pollution. *Adv. Aquat. Microbiol.* 1, 273, 1977.
5. Kaspar, C. W., Hartman, P. A., and Benson, A. K. Coagglutination and enzyme capture tests for detection of *Escherichia coli*, β-glucuronidase, and glutamate decarboxylase. *Appl. Environ. Microbiol.* 53(5), 1073, 1987.
6. Edberg, S. C. and Kontnick, C. M. Comparison of beta glucuronidase-based substrate systems for identification of *Escherichia coli*. *J. Clin. Microbiol.* 24(3), 368, 1986.
7. Edberg, S. C. and Edberg, M. M. A defined substrate technology for the enumeration of microbial indicators of environmental pollution. *Yale J. Biol. Med.* 61(5), 389, 1988.
8. Edberg, S. C., Allen, M. J., Smith, D. B., and the national collaborative study. National field evaluation of a defined substrate method for the simultaneous detection of total coliforms and *Escherichia coli* from drinking water: Comparison with presence-absence techniques. *Appl. Environ. Microbiol.* 55(4), 1003, 1989.
9. Atlas, R. M., Bej, A. K., McCarty, S., DiCesare, J., and Haff, L. *Monitoring Microbial Pathogens and Indicator Microorganisms in Water by Using Polymerase Chain Reaction and Gene Probes*. ASTM, Philadelphia (ASTM STP 1102), 1991.
10. Bej, A. K., Steffan, R. J., DiCesare, J., Haff, L., and Atlas, R. M. Detection of coliform bacteria in water by polymerase chain reaction and gene probes. *Appl. Environ. Microbiol.* 56(2), 307, 1990.
11. Bej, A. K., Mahbubani, M. H., Miller, R., DiCesare, J. L., Haff, L., and Atlas, R. M. Multiplex PCR amplification and immobilized capture probes for detection of bacterial pathogens and indicators in water. *Mol. Cell. Probes* 4(5), 353, 1990.
12. Bej, A. K., Mahbubani, M. H., and Atlas, R. M. Detection of viable *Legionella pneumophila* in water by polymerase chain reaction and gene probe methods. *Appl. Environ. Microbiol.* 57(2), 597, 1991.
13. Bej, A. K., McCarty, S. C., and Atlas, R. M. Detection of coliform bacteria and *Escherichia coli* by multiplex PCR: Comparison with defined substrates and plating methods for water quality monitoring. *Appl. Environ. Microbiol.* 57(8), 2429, 1991.

14. Bej, A. K., Mahbubani, M. H., DiCesare, J. L., and Atlas, R. M. PCR-gene probe detection of microorganisms using filter-concentrated samples. *Appl. Environ. Microbiol.* 57(12), 3529, 1991.
15. Cleuziat, P. and Robert-Baudony, J. Specific detection of *Escherichia coli* and *Shigella* species using fragments of genes coding for β-glucuronidase. *FEMS Lett.* 72(3), 315, 1990.
16. Mahbubani, M. H., Bej, A. K., Schaefer, F. W. III, Perlin, M. H., Jakubowski, W., and Atlas, R. M. Detection of *Giardia* using the polymerase chain reaction and distinguishing live from dead cysts. *Appl. Environ. Microbiol.* 57(12), 3456, 1991.
17. Mahbubani, M. H., Bej, A. K., Perlin, M. H., Schaefer, F. W. III, Jakubowski, W., and Atlas, R. M. The differentiation of *Giardia duodenalis* from other *Giardia* spp. based on the polymerase chain reaction and gene probes. *J. Clin. Microbiol.* 30(1), 74, 1992.
18. Bej, A. K., DiCesare, J. L., Haff, L. A., and Atlas, R. M. Detection of water-borne bacterial pathogens and indicators by polymerase chain reaction and gene probe methods using *uid*. *Appl. Environ. Microbiol.* 57(4), 101, 1991.
19. Mahbubani, M. H., Bej, A. K., Miller, R., Haff, L., DiCesare, J., and Atlas, R. M. Detection of Legionella with polymerase chain reaction and gene probe methods. *Mol. Cell. Probes* 4(3), 175, 1990.
20. De Leon, R., Baric, S., and Sobsky, M. D. Detection of enteric viruses by reverse transcriptase-polymerase chain reaction and non-radioactive oligoprobes. Abstracts of the 91st General Meeting of the American Society for Microbiology, Dallas, 1991.
21. Chang, G. W., Brill, J., and Lum, R. Proportion of β-glucuronidase-negative *Escherichia coli* in human fecal samples. *Appl. Environ. Microbiol.* 55(2), 335 , 1989.

CHAPTER 2

Allele-Specific Hybridization and Polymerase Chain Reaction (PCR) in Mutation Analysis: The *Salmonella typhimurium his* Paradigm

Thomas A. Cebula

PERSPECTIVES AND SUMMARY

Among test methods used in genetic toxicology, the *Salmonella*/Ames assay provides one of the most extensive data bases for evaluating the mutagenic/carcinogenic potential of a wide variety of chemicals, pharmaceuticals, and food additives. The utility of this test resides in its simple, relatively rapid, and reproducible endpoint: reversion of histidine auxotrophs to a wild-type phenotype (His⁻ → His⁺). The Ames test, as commonly employed, is straightforward. Portions of overnight nutrient broth cultures of particular *Salmonella typhimurium his* auxotrophs are treated with a suspected mutagen and plated; 2 days later, the resultant revertant colonies are counted. If mutant yields from treated cells are greater than twice that of untreated controls, the agent is considered mutagenic. Depending on the differential mutant yields obtained from the various tester strains, the potential of a chemical to act as either a frameshift or base-substitution mutagen (or both) can be quickly assessed.

Because the *Salmonella*/Ames assay is a reverse mutation system, one expects that certain mutations in the "targeted" genes will restore sufficient enzyme activity in the gene products to enable selection of His⁺ prototrophs. On the other hand, other types of mutations (for example, extended deletions and *IS* insertions) that can be isolated in forward test systems will be lost because of the selection criteria imposed in the *S. typhimurium* system. The latter two points emphasize the limits and bounds of this reverse mutation assay. Base substitution and frameshift mutations that lead to functional enzyme activity will be selected for, while mutations that do not restore enzyme activity will be lost in the selection process.

The quality of the *S. typhimurium* assay is judged by many standards, not the least of which is its sensitivity. Particular mutations placed in the background of the Ames strains have profoundly influenced the performance of this assay. For example, the presence of the *rfa* mutation genetically strips these strains of their lipopolysaccharide or endotoxin layer, thus removing a natural permeability barrier for many exogenously added agents. The deletion of *uvrB* disarms these strains of a major source of protection against DNA insults since the UvrB gene product is involved in normal *uvrABC* incision-excision repair of DNA. Moreover, the introduction of plasmid pKM101, which contains *mucAB* (a functional homologue of *umuDC*), makes these strains more prone to *recA*-dependent SOS error-prone repair. Finally, the use of microsomal extracts, which permits metabolic conversion to the ultimate mutagen/carcinogen, has further enhanced the sensitivity of these strains.

These modifications also afford mechanistic insights into how a particular chemical elicits its mutagenic response. For example, the role of incision-excision repair can be quickly assessed by comparing the induced mutant yields in the relatively isogenic *uvrB+* and Δ*uvrB* strains. The importance of SOS error-prone repair in the mutagenic process can be appraised by comparing mutant yields in isogenic strains with and without plasmid pKM101. The capacity of a mutagen to be direct-acting can be evaluated by the mutant yields obtained in strains with and without added microsomes. Further discussion of these refinements will be waived for they have been adequately reviewed elsewhere.[1-4]

This commentary will focus, rather, on the specificity of the *Salmonella*/Ames assay. Because the specificity hinges upon the particular *his* auxotrophic mutants used in the assay, the nucleotide sequences of four *his* mutations widely used in this assay—namely, *hisG46* (missense; C$\underline{T}$C → C$\underline{C}$C), *hisG428* (ochre; $\underline{C}$AA → $\underline{T}$AA), *hisC3076* (+1 frameshift; CCCC → CCCCC), and *hisD3052* (−1 frameshift; CCC → CC)—will be summarized, and the molecular alterations that lead to an His+ phenotype will be delineated. Moreover, the two direct means used in our studies, allele-specific colony hybridization and polymerase chain reaction (PCR) with direct DNA sequencing, will also be discussed.

At this writing, over 20,000 His prototrophs from *hisG46-*,[5-7] *hisG428-*,[6] *hisC3076-*,[8,9] and *hisD3052-*bearing[8,10,11] strains have been isolated and characterized in our laboratory. It should be emphasized at the outset that the *Salmonella*/Ames assay is attacked unfairly as being too limited by many, who errantly assume a priori that reversion means back mutation to true wild type. As this review will underscore, in addition to mutations to true wild type, there are multiple ways by which each of the mutational targets can be altered to yield pseudo–wild-type revertants. Indeed, the various furtive mutations revealed by our analyses negate such unwarranted attacks.

ALLELE-SPECIFIC COLONY HYBRIDIZATION

Because the mutations that lead to an His+ phenotype in these strains are localized to small clustered regions of known sequence, events can be readily

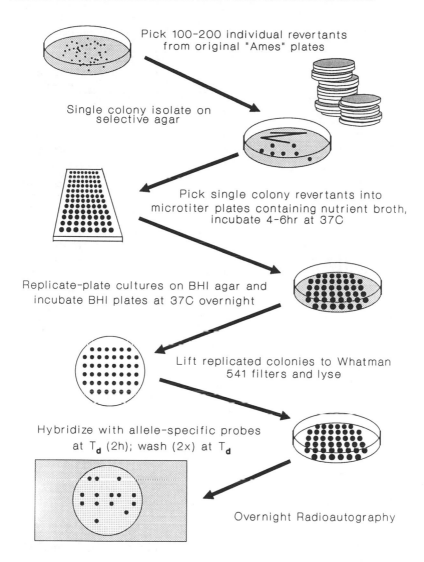

Pick 100-200 individual revertants
from original "Ames" plates

Single colony isolate on
selective agar

Pick single colony revertants into
microtiter plates containing nutrient broth,
incubate 4-6hr at 37C

Replicate-plate cultures on BHI agar and
incubate BHI plates at 37C overnight

Lift replicated colonies to Whatman
541 filters and lyse

Hybridize with allele-specific probes
at T_d (2h); wash (2x) at T_d

Overnight Radioautography

Figure 2.1 Schema for allele-specific colony hybridization.

identified using specific oligodeoxyribonucleotide (*n*-mer) probes. This technique takes advantage of the fact that stringent conditions can be defined where an *n*-mer will hybridize to its complementary sequence within denatured genomic DNA, while *n*-mers, differing by a single mismatch, will not.[12,13]

Protocol

The schema for allele-specific colony hybridization is depicted in Figure 2.1. Revertant colonies from an Ames pour plate assay are picked and purified on selective medium. Difco nutrient broth cultures (0.2 mL) of purified revertants

Table 2.1 Allele-Specific Probes Used in Mutation Analysis of *S. typhimurium*
his System

Probe	Event	Hybridization temperature (°C)	Washing conditions
G199-GTCGAT<u>CCC</u>GGTATT	—	47	3XSSC; 47°C
G199-GTCGAT<u>TCC</u>GGTATT	GC → AT	47	3XSSC; 47°C
G199-GTCGAT<u>CTC</u>GGTATT	GC → AT	47	3XSSC; 47°C
G199-GTCGAT<u>GCC</u>GGTATT	GC → CG	47	1XSSC; 50°C
G199-GTCGAT<u>ACC</u>GGTATT	GC → TA	47	3XSSC; 47°C
G199-GTCGAT<u>CAC</u>GGTATT	GC → TA	47	3XSSC; 47°C
5'-T<u>GGG</u>AAGGGTGAGGTCGG (tRNA[Thr3])	AT → CG[a]	60	6XSSC; 60°C
G610-CAGAGCAAG<u>TAA</u>GAGCTG	—	54	6XSSC; 54°C
G610-CAGAGCAAG<u>CAA</u>GAGCTG	AT → GC	56	6XSSC; 56°C
G610-CAGAGCAAG<u>TCA</u>GAGCTG	AT → CG	56	6XSSC; 56°C
G610-CAGAGCAAG<u>TAC</u>GAGCTG	AT → CG	54	6XSSC; 54°C
G610-CAGAGCAAG<u>GAA</u>GAGCTG	AT → CG	56	6XSSC; 56°C
G610-CAGAGCAAG<u>AAA</u>GAGCTG	AT → TA	54	6XSSC; 54°C
G610-CAGAGCAAG<u>TTA</u>GAGCTG	AT → TA	54	6XSSC; 54°C
G610-CAGAGCAAG<u>TAT</u>GAGCTG	AT → TA	54	6XSSC; 54°C
5'-TTTAATACCGGCATTCCC (*supB*)	GC → AT	52	6XSSC; 52°C
5'-ACTTTAAATCTGCCGTCA (*supC/M*)	GC → TA	50	6XSSC; 50°C
C724-GTGATCG<u>CCCC</u>TTATCCG	–C/G[b]	58	6XSSC; 58°C
C724-GTGATCΔCCCCCTTATCCG	–G/C[c]	58	6XSSC; 58°C
C731-CCCCCTTATCCGΔTTTCTACG	–C/G[d]	60	6XSSC; 60°C
C731-CCCCCTTATCCGCΔTTCTACG	–T/A[e]	60	6XSSC; 60°C
D874-CTGCCGCGΔCGGACACCGC	–CG/GC[f]	61	3XSSC; 61°C
D899-AGGC<u>C</u>CCTGAGCGCCAGT	+C/G[g]	60	3XSSC; 60°C

[a] Detection of an AT → CG transversion mutation in the tRNA[Thr3] locus. This probe is used to further characterize revertants that were shown to contain the original *hisG46* (CCC) mutation.

[b] Detection of *hisC3076* true wild-type revertants.

[c] Detection of *hisC3076* revertants containing a G/C deletion at C730.

[d] Detection of *hisC3076* revertants containing a C/G deletion at C742.

[e] Detection of *hisC3076* revertants containing a T/A deletion at C743.

[f] Detection of a CG/GC deletion in the alternating CG/GC octamer (D878–885). An unlabeled sequence (D892-GCCGGCAGGCCCTGAGCG) is also added with the labeled probe.

[g] Detection of +C addition in a run of three C residues (D902–904).

are propagated in microtiter dishes for 4–16 hr at 37°C. Individual cultures are simultaneously replica-plated, 44 at a time, to Difco brain heart infusion (BHI) agar plates. When the replica plating is completed, the individual microtiter cultures are fortified with dimethyl sulfoxide (DMSO) (20 μL) and the covered microtiter dishes stored at –70°C. These serve as working cultures for future studies of individual revertants.

Cultures transferred to BHI plates are then incubated overnight at 37°C. Colonies are "lifted" to Whatman 541 paper filters and lysed *in situ*.[5] After neutralization, filters are air dried at ambient temperature and used immediately in hybridization experiments or stored in a vacuum desiccator until needed. Filters stored up to 2 years have been used in hybridization experiments yielding results comparable to those obtained with freshly prepared filters.

Individual filters containing lysed revertants are placed in separate Falcon plastic petri dishes containing 10 mL of hybridization solution and equilibrated at the T_d (*vide infra*).[8] Labeled *n*-mer is added and incubation at the T_d

continued for 2 hr in a New Brunswick Scientific G24 shaker with agitation (75 rpm). The filters are subsequently subjected to two 20 min washes with 1–6 X SSC (SSC is 0.15 M NaCl and 0.015 M sodium citrate, pH 7.0) at the T_d. The filters are air dried and exposed to Kodak® XAR-2 film for 2–18 hr at –70°C in the presence of Quanta III (DuPont®) intensifier screens. Hybridization and washing conditions are summarized in Table 2.1; exact details for particular *his* probes have been published elsewhere.[5–6,10]

Perhaps the most tedious step in this procedure is the isolation and purification of the *his* revertants from the original Ames experiment. After having an ordered array of revertants on BHI plates, the remaining steps (lysis, hybridization, washing, and development of X-ray film) often can be accomplished within a single day. In using this technique, a number of considerations early on might help ensure success.

Heteroduplex Formation

The specificity and sensitivity of allele-specific colony hybridization are ultimately dictated by Watson–Crick base pairing of "probe" DNA and "target" DNA. Studies of heteroduplexes in solution have demonstrated that the replacement of a G/C or A/T base pair by any mismatch results in reduced thermal stability of the helix.[14,15] However, because particular mismatches are capable of stacking within the helix, the resultant heteroduplexes can be quite stable. The reader is referred to the Modrich article[14] and references therein for a very good discussion of the pairing of rare base tautomers, wobble pairs, ionized base forms, and *anti–syn* isomers. Hierarchal relationships for the various mismatches have been derived, though, as one might expect, the stability of a particular mismatch is profoundly influenced by the local sequence environment.[14,15]

For example, Werntges et al.[15] examined the 16 possible combinations of A, T, C, and G at a particular site, X/X′, within a synthetic heteroduplex 18 nucleotides in length (dCGTCGTTT X ACAACGTCG·dCGACGTTGT X′ AAACGACG). They found the stabilities of the various base pairs to be [G/C > C/G > A/T > T/A] > [G/T = T/G] > A/G > [T/T = C/T] > [G/G = G/A] > T/C > A/A > A/C > C/A > C/C. Whereas the T_m values for heteroduplexes containing correct Watson–Crick base pairs (64.9–69.6°C) were significantly greater than the T_m values for heteroduplexes containing mismatches (51–60°C), the shapes of the temperature curves for particular mismatches, led Werntges et al. to conclude that the transition from helix to random coil was not an all-or-none process. Rather, the stacking potential of nearest neighbors and next-nearest neighbors contributed to the overall stability of the heteroduplex.[15] Indicative of these long-range interactions was the nonidentity of T_m values obtained for heteroduplexes containing G/C (69.6°C) and C/G (68.1°C) or A/T (67.1°C) and T/A (64.9°C) base pairs. As with Watson–Crick base pairs, particular mismatches, e.g., A/G (59.3°C) and G/A (56.6°C), showed context effects, while others, e.g., G/T (60.1°C) and T/G (60.0°C), did not.

Wallace et al.,[12] in examining hybridization of synthetic oligonucleotides to DNA affixed to a filter substratum, defined the parameter T_d as the temperature at which half of the n-mer becomes dissociated from the filter-bound target DNA; although not as rigorously measured, the T_d should approximate the T_m value for "helix to random coil" transitions determined in solution. For duplexes 11–23 base pairs (bp) long in 1 M Na$^+$,

$$T_d = 2° \times (\text{number of A + T}) + 4° \times (\text{number of G + C})^{16,17} \qquad (1)$$

The calculated T_d for each of the *his* probes has served well in helping to establish our hybridization (and washing conditions) for *S. typhimurium* revertants.[5,6,10] By carrying out hybridizations at or near the T_d, one can basically eliminate the formation of mismatched duplexes without affecting the formation of the correct duplex. Moreover, at the higher temperatures, unspecific binding of the labeled probe to the filter substratum is reduced.

Probe Considerations

Synthesis and Labeling of Synthetic Probes

The necessary *his* probes (and primers) for our analyses were synthesized by phosporamidite chemistry using an Applied Biosystems 380B DNA synthesizer (Applied Biosystems, Foster City, CA). Purification was accomplished using Applied Biosystems reverse-phase purification cartridges following conditions provided by the manufacturer. T_4 polynucleotide kinase (ATP: 5′-dephosphopolynucleotide 5′-phosphotransferase; EC 2.7.1.78, New England Biolabs, Beverly, MA) and [λ-^{32}P]ATP (>7000 Ci/mmol, ICN Radiochemicals, Irvine, CA) were used to 5′-end label the various n-mers for allele-specific colony hybridization experiments.[5,10] Labeled probe was resolved from unincorporated label using either NACS columns (Bethesda Research Laboratories, Gaithersburg, MD) or NucTrap™ push columns (Stratagene, La Jolla, CA). Although nonradioactive approaches have been tried and, indeed, were successful in a dot-blot format,[18] we have been unable to adapt these techniques to a colony hybridization protocol.

Probe Length

The length of the n-mer dictates how discriminating a probe will be in identifying unique sequences, and this will depend upon the complexity of the genome under investigation. Assuming an equal distribution of four nucleotides randomly distributed in a genome size of N nucleotide pairs, the length of a unique sequence, n, will be defined by

$$n = \log 2N/\log 4 \qquad (2)$$

Thus, for the 5×10^6 bp genome of *S. typhimurium*, which is free of highly repetitive DNA and has a G/C content of about 52%, a sequence length of $\geq$12 nucleotides should be unique. However, since n-mer length is proportional to duplex stability, in practice, longer n-mers are usually used to ensure a more rugged screening procedure. This is somewhat of a compromise for although the rate of formation of perfectly matched and mismatched duplexes may be independent of n-mer length, the rate of dissociation of mismatched duplexes appears to depend on n-mer length (11-mer > 14-mer > 17-mer).[12] One should note that from the design of the Wallace experiment,[12] it is hard to judge whether it is the position of the mismatch or the length of the n-mer that accounts for these differences.

In our studies, we use routinely 15-mers to analyze $hisG46$ revertants,[5-7] 18-mers to examine $hisG428$[6] and $hisD3052$[10] revertants, and 18- to 20-mers to probe $hisC3076$ revertants.[9] The n-mer are designed usually with the crucial deoxyribonucleotides at or near the center of the n-mer to maximize the effects of cooperative melting of mismatched duplexes. Although this strategy has proved sound in the majority of cases, we have found for certain 18-mers that better discrimination is obtained when the mismatch is 4–5 bases from either the 5'- or 3'-end.

Palindromy

When designing n-mers as specific probes, sequences containing inverted repeats should be avoided when possible. The self-complementarity of such sequences could affect 5'-labeling[17] and might affect heteroduplex formation. Sometimes such sequences are unavoidable. This is especially true when applying this technique to the analysis of mutations. Since palindromic or near-palindromic sequences are mutational hot spots, perhaps because of the secondary structures such sequences can adopt, the appropriate n-mer might indeed contain an inverted repeat.

Kupchella and Cebula,[10] for example, encountered this difficulty in the analysis of $hisD3052$ revertants. A frequent reversion event in $hisD3052$-containing strains is the loss of 2 bp (CG/GC) from an alternating run of CG base pairs. The probe designed to detect these deletions is comprised of the following sequence: 5'D874-C CTGCCGC^GCGGACA CCGC. Note the near-perfect palindrome (underlined text) contained within this sequence. Furthermore, as will be discussed in greater detail below, the $hisD3052$ target DNA is rich in short, inverted repeats that could facilitate formation of secondary structure within this sequence. When used in hybridization experiments, the probe yielded a specific, though meager, signal. We reasoned that the potential for both the probe *and* target DNA to form hairpin loops was confounding the hybridization analysis. Therefore, an unlabeled n-mer (5'D892-GCCGGCAGGCCCTGAGCG) was added along with the labeled probe to competitively melt out any secondary structures. This strategy resulted in a dramatic improvement in the performance of the CG/GC deletion probe.[10] Thus, competitive hybridization offers one potential solution if one happens to encounter a "recalcitrant" probe. However, it is not a sure cure-all for this problem since (1) we have utilized other sequences containing inverted repeats successfully without the need of competitors, and (2) particular probes remain refractory to this approach.

Hybridization Times

It should be emphasized that, for convenience, colony hybridization experiments are usually carried out for extended periods of time, i.e., overnight (>16 hr) incubations. As we[8,11] and others[17] have stressed, since colony hybridization follows pseudo-first-order kinetics and the labeled n-mer is in excess to target DNA, equilibrium is achieved quite rapidly (tens of minutes to a couple of hours, depending upon the concentrations of probe and target DNA and the experimental conditions). If hybridizations are conducted for prolonged periods of time, the chances of nonspecific and unspecific binding are markedly increased. Thus, we have found for particular *his* probes, especially those that are G/C rich, that shorter hybridization times (2 hr) have allowed maximum discrimination of "correct" and mismatched sequences.

Probe Refinement

Certain approaches may be tried to improve the discriminating power of a potential probe. Due to end effects, the stability of a "probe"-"target" heteroduplex will be defined by the n-mer sequence, i.e., the shorter of the two DNA molecules. As a general rule, one usually tries to design a probe that has approximately a 50% G/C content, yet, as seen in Table 2.1, depending on the mutational target, this may not always be feasible. A perusal of Table 2.1 shows that n-mers, ranging in G/C content from 50 to 83%, have been used with success for the various *his* mutation targets.

The relative stability of a particular heteroduplex can be manipulated in a number of ways. These include adjusting the G/C content, increasing or decreasing the length of the n-mer, or applying more stringent hybridization conditions such as increasing the temperature and/or decreasing the Na^+ concentration. If a particular probe is not discriminating enough, perhaps one might be able to use its complement to alleviate this problem. Alternatively, shifting the sequence context slightly might dramatically improve resolution. Various software packages are now available to help choose appropriate n-mers as probes (or primers). Although they are very valuable in guiding the design of these useful reagents, the reader should be mindful that the exact rules governing the technique have not, as yet, been written. Before the definitive software is designed, it is clear that many experiments must still be performed to unmask the subtle nuances of DNA "probery."

POLYMERASE CHAIN REACTION (PCR)/DIRECT SEQUENCING

Protocol

The schema for PCR amplification and DNA sequencing is shown in Figure 2.2. Revertants not identified with allele-specific probes are grown overnight in Difco nutrient broth. Portions (1 mL) of these cultures are washed and resus-

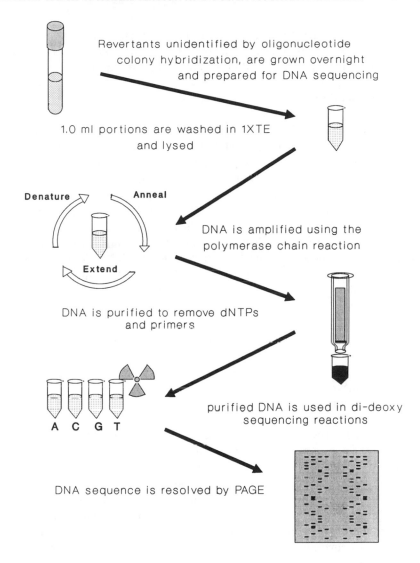

Revertants unidentified by oligonucleotide
colony hybridization, are grown overnight
and prepared for DNA sequencing

1.0 ml portions are washed in 1XTE
and lysed

Denature   Anneal

Extend

DNA is amplified using the
polymerase chain reaction

DNA is purified to remove dNTPs
and primers

A   C   G   T

purified DNA is used in di-deoxy
sequencing reactions

DNA sequence is resolved by PAGE

Figure 2.2 Schema for PCR and direct sequencing.

pended in 0.5 mL TE buffer (10 mM Tris-HCl, 1 mM EDTA, pH 7.4). The resuspended cells are subjected to freeze–thaw lysis,[10] and DNAs (10–100 ng) of the resultant lysates are used as templates for amplification by automated PCR. The amplified DNA fragments from PCR reactions are purified using either Sephadex G-25 spun columns[19] or Millipore Ultrafree-MC filters with a MW cutoff of 30,000. The eluates from the columns are ethanol precipitated and resuspended in deionized H_2O. Dideoxy sequencing[20] of the various purified PCR products are conducted using [α-^{35}S]dATP (>1000 Ci/mmol; Amersham, Clearbrook, IL) and Sequenase® (U.S. Biochemicals, Cleveland, OH); the sequences are resolved using 6% denaturing polyacrylamide gels. In sequencing *hisD3052* DNA, both dGTP and dITP reactions are simultaneously run to circumvent compression artifacts.[10] Since compressions were not observed in the other targets, the dITP reactions were omitted.

PCR Amplification

Since its original description in 1985,[21] PCR has proved to be a versatile and integral technique, one that has revolutionized experimental approaches in the various areas of biology. PCR is elegant in its simplicity, providing a rapid means to amplify specific DNA segments selectively from either a prokaryotic or eukaryotic genome. The intrinsic enzymatic properties of DNA-dependent DNA polymerases make PCR possible. That is, although DNA polymerases cannot *initiate* DNA synthesis, they can efficiently catalyze a Mg^{2+}-dependent and primer-directed chain extension reaction at the expense of the four deoxyribonucleotide substrates (dNTPs = dGTP, dCTP, dATP, and TTP) when supplied with appropriate primers and template DNA. Synthesis proceeds unidirectionally ($5' \rightarrow 3'$) yielding a strand of DNA that is complementary to the template DNA.

The use of the highly processive, heat-stable DNA polymerase from *Thermus aquaticus* (*Taq* polymerase) makes feasible automated PCR. Three separate processes occur within each cycle of PCR: denaturation of the double-stranded DNA template, annealing of unique primers to sequences flanking the target, and chain elongation ($5' \rightarrow 3'$) of the primers across the targeted DNA region. Because concentrations of enzyme, dNTPs, and primers are not rate limiting, each PCR cycle results in a near doubling of the desired DNA sequence. Thus, at the end of 20–30 cycles of the PCR, in theory, a particular DNA sequence can be selectively amplified a million- to a billion-fold.

Ultimate yields will vary, of course, depending upon the instrumentation employed, the specific experimental conditions, the particular sequence being amplified, and the primer sets used for amplification. The exact conditions for PCR amplification will also differ, depending on whether double-stranded or single-stranded sequencing techniques are utilized to characterize the amplified product.[10] Moreover, it is especially important to note that since there are now various commercial sources for *Taq* polymerase and other thermostable DNA polymerases as well, the yields may vary considerably with the enzyme and/or the batch (and formulation) of enzyme used to amplify the intended DNA fragment. In our studies, *Amplitaq®* (Perkin-Elmer Cetus, Norwalk, CT) was used exclusively to amplify the *hisD3052* target. For the other targets, *Taq* polymerase from Digene Diagnostics Inc. (Beltsville, MD) was employed.

PCR Conditions

For amplification of the *his* targets, master mixes were prepared according to the *Amplitaq®* protocol supplied by the manufacturer. The final volume of each PCR reaction mix was usually 100 µL and contained 10 m*M* Tris-HCl (pH

8.3), 50 mM KCl, 1.5 mM MgCl$_2$, 200 µM of each of the dNTPs, portions of the clarified lysates (vide supra), primers, and 3.5 units of *Taq* polymerase. The reaction mixes were overlayed with 50 µL of mineral oil, and PCR cycling commenced immediately. When single-stranded product was required, the asymmetric amplification protocol described by Gyllensten and Erlich[22] was utilized; primer sets for each of the targets were used such that one of each primer set was in 50-fold molar excess of the other. If double-stranded product was desired, the conditions were the same, except that equimolar amounts (50 pmol) of each primer were added to the reaction mix. Exact PCR cycling conditions and times varied depending upon the *his* target being amplified. In brief, a DNA thermocycler (Perkin Elmer-Cetus, Norwalk, CT), programmed for 35 step cycles of 1.5 min at 94°C (denaturation), 1–1.5 min at the T_m (annealing), and 1–2 min at 72°C (chain elongation), was used for amplification. After amplification, 20 µL of each PCR sample was analyzed by electrophoresis at 100–150 V for 1.5 hr in 1.8% agarose gel in Tris-borate-EDTA[19] containing ethidium bromide (0.2 µg/mL). A 123-bp ladder (Bethesda Research Laboratories, Gaithersburg, MD) was used as the molecular size standard. When single-stranded sequencing was used, two 100 µL PCR reactions were pooled and purified for sequencing reactions.

Primers are designated 5′ →3′ relative to the first base of each start codon of the particular gene. For asymmetric amplification, the limiting primer is denoted by an asterisk. GCCTGATTGCGATGGCGG (G113–130) and GTCAAGACGGCGCAGGG* (G284–300C) were used to amplify a 188-bp fragment encompassing the *hisG46* locus. For *hisG428*, GCGACGCTTGAAG-CTAACG (G523–541) and CTTGGCGCGTGCATCATGA* (G686–704C) were utilized for amplification of a 182-bp portion of *hisG*. For amplification of a 343-bp portion containing the *hisD3052* allele, GCCGGGCCGTCTGAAGTACTG* (D700–720) and CCGAGCCTGCGCTGGTAATCG (D1022–1042C) were employed. In each of these three cases, the PCR primers served subsequently as sequencing primers. The amplification of a 490-bp fragment containing *hisC3076* was accomplished using TTGGCGTAGAGCGCCGGACGG* (C368–388) and GGGCATTCACCAGATACTGACG (C835–856C). For sequencing, an internal primer, GCTGGCGGGTCTGCGCTG (C660–677), was utilized.

MUTATIONAL TARGETS

hisG46

Mutant *hisG46* was originally isolated as a spontaneous *his* auxotroph from *S. typhimurium* LT-2.[23] Sequence analysis[24] showed that an A/T → G/C transition mutation occurred at *hisG* codon 69 (residues 433–435 of the *his* operon), which leads, by virtue of the substitution of Pro (CCC) for Leu (CTC), to a nonfunctional HisG product (1-[5′-phosphoribosyl]-ATP: pyrophosphate phosphoribosyltransferase; EC 2.4.2.17). *hisG46*-bearing strains are reverted

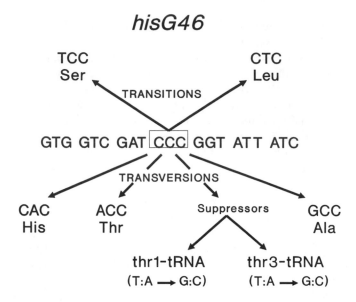

Figure 2.3 *hisG46* target sequence and mutations that lead to His⁺ phenotype.

readily by a wide variety of base-substitution mutagens. Figure 2.3 shows the sequence context of the *hisG46* allele and details seven major ways that an His⁺ phenotype can be restored. Single base pair transition or transversion mutations at either of the first two positions of the *hisG46* codon can give rise to His⁺ prototrophs, whereas mutations at the third position will not be recovered since they are silent (i.e., CCX where X = A, C, G, and T encode Pro). G/C →C/G transversion events at the second position of the codon (Pro → Arg) are not recovered (not shown in Fig. 2.3), unless the *hisG46* allele is carried on a multiple-copy plasmid.[25]

   Reversion of *hisG46* strains can also occur via intergenic suppression, i.e., mutations at extragenic sites. Levin and Ames[26] presumed these to be A/T → C/G transversion mutations in unlinked tRNA genes, based on the known chemistry of angelicin and its ability to induce extragenic suppressors in *hisG46*-bearing strains. Following their cues, we reasoned that the most likely possibilities for these suppressors were mutant tRNAᵀʰʳ products (GGU → GGG) as transversions at the other anticodon positions would lead either to synonymous changes within tRNAᴾʳᵒ (UGG → GGG) or to changes in the unique tRNAᴴⁱˢ species (GUG → GGG). Indeed, we recently identified two *hisG46* suppressor loci as mutant alleles of the tRNAᵀʰʳ¹ and tRNAᵀʰʳ³ genes,[27] A/T →C/G mutations in the anticodons of these tRNAs presumably suppress the *hisG46* mutation by inserting thr, rather than Pro, at codon 69 of the HisG product.

   Four of the six possible base pair substitutions can be monitored with the *hisG46* system; G/C → A/T transitions and G/C → T/A transversions at each of two intragenic sites, G/C → C/G transversions at one intragenic site, and A/T → C/G transversions at two extragenic sites (see Fig. 2.3). Because of the finite ways that *hisG46*-bearing strains can revert to His$^+$, this target lends itself to exploration by allele-specific colony hybridization. Six specific oligodeoxyribonucleotides can be used to identify virtually 100% of the *hisG46* revertants.

The usefulness of allele specific hybridization was emphasized by Miller and Barnes[28] when they showed that 5-azacytidine, which gave rise to a nominal increase in mutant yields, enhanced G/C → C/G transversion events about 8-fold at this locus. Similarly, we underscored the effectiveness of this technique in examining photoactivated psoralen (PUVA)-induced *hisG46* revertants. That is, PUVA induced about a 14-fold increase in suppressor mutations (approximately a 30-fold increase if mutant yield were corrected for the contributions of spontaneous mutations) even under conditions where mutant yields in induced populations did not exceed twice that of controls.[5]

We have further exploited allele-specific DNA hybridization to examine the mutational specificity of *N*-methylnitrosourea (MNU), nitrosoguanidine (MNNG), methyl methanesufonate (MMS), sodium azide, 4-nitroquinoline oxide (NQO), benzo[*a*]pyrene (BP), nitrofurantoin (NF), and aflatoxin B$_1$ (AFB$_1$) in *S. typhimurium* strain TA100 (*rfa, hisG46,* Δ*uvrB*/pKM101).[7] These eight mutagens produced four unique classes of reversion spectra that differed from the spontaneous,[7] ultraviolet light (UV)-,[29] PUVA-,[5] γ-ray-,[29] and 5-azacytidine-induced[28] spectra.

For example, the vast majority of MNU- (>90%), MNNG- (>90%), and sodium azide-induced revertants (>80%) in strain TA100 were the result of transition mutations. Quite surprising, however, was the site bias exhibited by these agents. Whereas both MNU and MNNG primarily induced transition mutations at site one (80%; CCC → TCC), virtually all (99%) of the sodium azide-induced mutations analyzed occurred at site two (CCC → CTC).[7] Mutations induced by a nitric oxide-generating compound were recently characterized in a *hisG46*-bearing strain.[30] Consistent with the deaminative properties of nitric oxide, all mutations recovered were transition mutations; a strong bias for site two was also noted in these studies.[30] UV-induced transition mutations, in contrast, were equally distributed between the two selectable sites.[29]

In contrast, primarily CA transversion mutations were induced by MMS (74%), NQO (72%), BP (70%), NF (78%), and AFB$_1$ (85%).[7] Both NF and AFB$_1$ exhibited a strong site preference at the *hisG46* locus. That is, greater than 90% of the C → A transversion events induced by these compounds occurred at site two (CCC → CAC). This site bias, though less pronounced, was also observed in MMS-, NQO-, and BP-induced revertants. Here, site two revertants were recovered approximately two times more frequently than were site one (CCC → ACC) revertants.

Thus, of 13 mutagens examined in *hisG46*-bearing strains, eight unique mutational spectra, distinct from spontaneous, have been obtained, emphasizing the discriminating potential of this simple mutational target. In our analyses of approximately 3000 *hisG46* revertants, about 1% of the mutagen-induced revertants, were refractory to our probing analysis. PCR amplification and direct sequencing of probe-negative revertants revealed that these mutants

hisG428

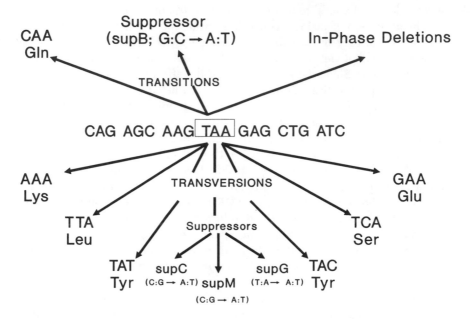

Figure 2.4 *hisG428* target sequence and mutations that lead to His⁺ phenotype.

contained multiple base pair substitutions at or near the *hisG46* locus;[7] these included tandem double and triple mutations within the *hisG46* codon and vicinal double mutations.

hisG428

Mutant *hisG428* was isolated from an *ara-9* strain of S. *typhimurium* LT-2 treated with 2-aminopurine;[23] *hisG428* was classed as an ochre mutation and shown to map to the distal region of the *hisG* gene.[23,31] Sequence analysis[32,33] established that a G/C →A/T transition mutation had occurred in codon 207 (CAA → TAA) of the *hisG* gene. Strains harboring *hisG428* are reverted by a variety of oxidative mutagens. Figure 2.4 shows the sequence context of the *hisG428* ochre mutation. Note that two of the nine possible base substitutions within the ochre codon will generate other nonsense codons (TAA → TGA; TAA → TAG) and thus will be lost in the selection procedure. The other seven, however, might be expected to yield an His⁺ phenotype. Indeed, using allele-specific DNA hybridization, we have identified each of the seven intragenic base substitutions[6] (Figure 2.4). Four of the base pair substitutions identified in our analysis [TCA(Ser), GAA(Glu), TAT(Tyr), and TAC(Tyr)] had not been previously reported among *hisG428* revertants.[26]

Reversion of *hisG428* strains is somewhat more complex than that described for the *hisG46* system. In addition to the seven intragenic events, four extragenic

Table 2.2 **Mutations That Yield an His⁺ Phenotype in the *hisG46* and *hisG428* Systems**

Event	*hisG46*	*hisG428*
G/C → A/T	TCC(Ser), CTC(Leu)	*supB*
A/T → G/C	—	CAA(Gln)
G/C → C/G	GCC(Ala)	—
G/C → T/A	ACC(Thr), CAC(His)	*supC*, *supM*
A/T → C/G	*thr-1tRNA, thr-3tRNA*	GAA(Glu), TAC(Tyr), TCA(Ser)
A/T → T/A	—	AAA(Lys), TTA(Leu), TAT(Tyr), *supG*

loci can suppress the *hisG428* mutation (Figure 2.4). Because extragenic events represent a major proportion of spontaneous *hisG428* revertants,[6,32,33] we have designed specific oligonucleotides to identify each of the suppressor loci.[6] This approach offers a facile alternative to the more cumbersome genetic mapping of these mutants.[26] In-phase deletions are yet another means by which an His⁺ phenotype can be restored[33] in *hisG428*-bearing strains. Direct sequencing of probe-negative *hisG428* revertants from our collection has revealed deletions ranging in size from 3 to 12 bp that encompass part or all of the ochre codon.

The *hisG428* and *hisG46* systems are complementary; between the two, each of the 6-bp substitutions can be effectively monitored (Table 2.2). Used together, these mutational targets should provide puissant insights into the mechanisms by which particular mutations arise. Recently, for example, Prival and Cebula[6] used these targets to investigate the mechanisms underlying the phenomenon known as "directed mutation." Likewise, Guttenplan et al.[34] used both targets to explore the effects of 2-hydroxylation on the mutagenic activity and specificity of alkyldiazonium ions. The combined analysis of *hisG46-* and *hisG428*-bearing strains pays yet another dividend. Since, in four of the six cases, the events can be monitored at more than one locus (Table 2.2), the role that sequence context plays in mutation formation can be readily assessed under the same selective pressures.

hisC3076

Mutant *hisC3076* was isolated from an *ara-9* strain of *S. typhimurium* LT-2 after treatment with ICR364-OH.[35] The *hisC3076* mutation was recently sequenced[8] and shown to be a one base pair addition (C/G) in a run of four C/G base pairs (bp C731–734; Fig. 2.5). Strains containing *hisC3076* lack a functional histidinol-phosphate aminotransferase (EC 2.6.1.9) and thus require histidine for growth. A wide variety of agents, including some that preferentially react with A/T base pairs, readily revert *hisC3076*-containing strains.

From an analysis of approximately 1600 *hisC3076* revertants arising spontaneously or induced with MNNG, AFB$_1$, BP, 9-aminoacridine, UV, PUVA, adriamycin, and daunomycin, Cebula and Koch[11] concluded that the mutation target for intragenic suppression of *hisC3076* was approximately 20 bp. That is, although the primary reversion event was the loss of a C/G base pair which restored true wild-type sequence (50–100%, depending upon the mutagen),

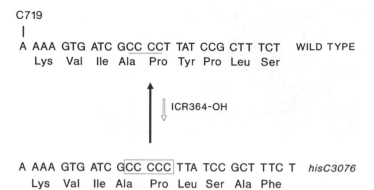

Figure 2.5 *hisC3076* target sequence and mutations that lead to His+ phenotype.

virtually any base pair along a 21-bp tract of the mutant sequence (bp C725–745) can be deleted to yield a pseudo-wild-type revertant.[9]

Note that selection precludes the recovery of minus one events 5′ to bp C725 because such deletions will generate an amber codon (Fig. 2.5; bp C725–727). The reason why minus one deletions are not recovered 3′ to bp C745 is not so clear since the next nonsense codon that would be generated is approximately 150 bp downstream from this site. Presumably, the 3′-distal portion of the *hisC3076* target must encode an essential domain of the HisC product; thus, mutations in this region would be lost in the selection procedure.

Three unique plus two events have also been recovered along this same stretch, though it appears that plus corrections of this allele represent a minority class of reversion events.[9] Complex mutations (base substitution *and* a shift in reading frame) are also frequently recovered among *hisC3076* revertants (2–36%, depending upon mutagen treatment); these mutations are exclusively due to the inducible SOS repair pathway.[35]

hisD3052

Mutant *hisD3052* was isolated from an ICR364-OH-treated *ara-9* strain of *S. typhimurium* LT-2.[36] In 1974, Isono and Yourno[37] deduced from polypeptide analysis that the *hisD3052* mutation was the result of a C/G deletion in a run of three C/G base pairs (bp D893–895) of the *hisD* gene. Nucleotide sequence data, more than a decade later,[4] confirmed these earlier results. This deletion results in the generation of several nonsense codons in close proximity of the *hisD3052* site (bp D905–907, D920–922, D926–928, D938–940; the base number assignments refer to the wild-type sequence). Strains harboring *hisD3052* lack a functional histidinol dehydrogenase (EC 1.1.1.23), the enzyme that catalyzes ultimate conversion of histidinol to histidine. Sequences surrounding the *hisD3052* mutation are G/C rich and contain a number of direct and inverted repeats. Inverted repeats lend themselves to quasipalindromic structures such as the two examples depicted in Figure 2.6A and C.

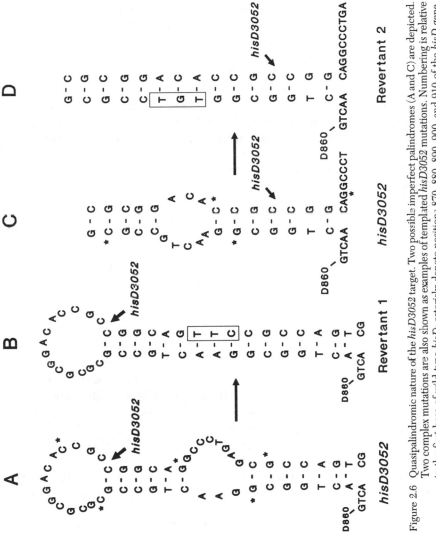

Figure 2.6 Quasipalindromic nature of the *hisD3052* target. Two possible imperfect palindromes (A and C) are depicted. Two complex mutations are also shown as examples of templated *hisD3052* mutations. Numbering is relative to the first base of wild-type *hisD*; asterisks denote positions 870, 880, 890, 900, and 910 of the *hisD* gene.

As reasoned by Isono and Yourno[37] and confirmed by allele-specific colony hybridization experiments[10,38] and sequencing,[39-41] the predominant reversion event in both spontaneous and mutagen-induced revertants is the loss of 2 bp (CG/GC) from an alternating CG (GC) octamer 5′ to the *hisD3052* allele (bp D878–885; Fig. 2.6). The design of a specific probe to identify these particular CG/GC deletions[38] has permitted rapid characterization of other sequence alterations in the *hisD3052* target by PCR and direct sequencing.[8,10,11,42] Unlike the *hisC3076* system, reversion to wild type of *hisD3052*-bearing strains is remarkably rare. Rather, virtually all (99%) revertants are pseudo-wild type; additions ($3n + 1$) and deletions ($3n - 1$) ranging in size from 1 to 34 and 2 to 23 bp, respectively, are responsible for the His⁺ phenotype. For example, in our collection of more than 2000 spontaneous *hisD3052* revertants, 130 distinct events that impart an His⁺ phenotype have thus far been characterized. This figure, by our estimates, represents approximately 70% saturation of the *hisD3052* target.

At present, the most extreme 5′-single bp addition is a T/A at position D854; this we define as the 5′-boundary of the *hisD3052* target. The 3′-boundary is specified by the third position of the first TGA codon since single base pair additions 3′ to this ultimate site will not be recovered. Thus, for plus one events, the *hisD3052* target is 54 nucleotide bp. Within this nucleotide tract, we have recovered at least 26 different 1-bp insertions that lead to an His⁺ phenotype. In this region of *hisD*, there are 12 positions where a particular base is vicinally reiterated. Such sites might be more prone to Streisinger slippage.[43] Consistent with this idea, single base pair additions were recovered at each of these 12 sites. At each of these sites, the addition was a further iteration of the repeated base. Furthermore, most of the multi-base-pair additions in our collection can be explained by a modified Streisinger mechanism since they represent partial duplications of sequences immediately adjacent to the "inserted" sequence.

Single-base-pair additions also have been recovered at nine different sites that do not contain iterative runs of bases. In each case, the mutation completes a repeat of a neighboring sequence. We give here three examples of these mutations. For clarity, the mutation is presented in lower case, flanking sequences in upper case, and the direct repeat is underlined. One +G mutation was recovered at D877, and the flanking sequence was AACTGGCGGAACTGgCC. The +A mutation at D883 has as its sequence context TGCCGCGCaGCGGACACCG-CCGGCAG. The sequence context of the +A mutation at D888 is GCGGAACTGCCGCGCGCGGAaC. Sequences of 7, 9, and 7 bp in these three cases, respectively, appear to presage the type of plus one event recovered, suggesting that nearby sequences may be templating these mutations.

Over 70 unique deletions have also been characterized in our collection; these deletions involve a 63-bp (D855–918) tract of the mutant *hisD* gene. As one might expect, the target size is somewhat larger than that defined for $3n + 1$ insertions since $3n - 1$ deletions that encompass the first nonsense codon (D905–907) and extend 3′ will relieve polarity. Classical Streisinger slippage can account for some of these deletions. For example, the major minus two deletion,

CG (or GC) at bp D878–885, can be explained by slipped mispairing within the recurring CG octamer. Since 14 different "slips" within the octamer can lead to this mutational outcome, this replication-dependent model may indeed rationalize why the –CG (GC) event is so frequently recovered.

Nineteen other unique minus two events are represented in our spontaneous collection. Many of these particular frameshift errors might be better explained by the transient misalignment model proposed by Kunkel and colleagues.[44,45] This model predicts that specific base-substitution mutations will occur within unusual sequence environments and that such mutations will induce slipped mispairing. From this model, one might expect, if selection is unbiased, that in the minus one frameshift hisD3052 system "paired" revertants should be recovered. Indeed, we find, for example, a transversion mutation (G → C) at D886 could explain both the GG (D885–886) and GA (D886–887) deletions found in our analyses. Similarly, a C → G transversion at D890 would account for both the CC (D890–891) and GC (D892–893) deletions that we have characterized.

The importance of short, direct repeats in deletion formation has been emphasized;[46] the size of the deletion is defined by the length of the intervening sequence plus one copy of the direct repeat. In our spontaneous collection, deletions of 14 bp (D901–914; CAGgccctgagcgccag), 17 bp (D896–912; CGCCggcaggccctgagcgcc), and 11 bp (D871–881; GCGgaactgccgcg) are but three examples where deletion endpoints may be delimited by such repeats. Many more deletions within our spontaneous collection might similarly be explained by "near-perfect" repeats, though one wonders if more subtle cues provoke these mutations.

Finally, the use of PCR and direct sequencing analysis has uncovered an intriguing array of complex mutations in hisD3052 revertants. Complex mutations fall into two classes. Class I events are most likely due to SOS-processing since these events are recovered more frequently in pKM101-containing strains than in strains lacking this plasmid. Moreover, mitomycin, a strong SOS-inducing agent, induces virtually all (>90%) Class I events.[47] The signature for such events is a base substitution mutation and a concomitant frameshift mutation, either a plus one addition or a minus two deletion. Noteworthy is our finding that in mitomycin-induced revertants, complex minus two deletions are confined to the CG octamer whereas complex plus one additions occur both in this pocket as well as flanking sequences.[47] Class I complex mutations were also recovered in ellipticine-induced hisD3052 revertants.[42]

The hallmark of Class II complex mutations is the seemingly random base pair additions and accompanying deletions that bring hisD back into a proper reading frame. Though experiments are still in progress, it is unlikely that such events are the product of SOS-processing since Class II complex mutations were recovered in equal proportions from hisD3052 strains with or without plasmid pKM101. A careful inspection of Class II events commands a more disciplined interpretation of such mutations.

For example, Kupchella and Cebula[10] discussed the palindromic nature of a +34-bp "insertion" near the *hisD3052* site and how this mutation probably arose by fold-back synthesis.[48] This illicit replication/repair process, ordinarily proscribed, would be expected to generate inverted repeats within the target sequence. Two other mutations detailed below and depicted in Figure 2.6B and D might also be explained by this process. A +TTC mutation at D901 accompanied by the deletion of GCCCTGAG (D901–908), for example, could be explained by fold-back synthesis templated by a neighboring sequence:

D864<u>ACTGGCGGAACTGCC</u>GCGCGCGGACACCGC<u>CGGCAG</u>ttc<u>CGCCAGT</u>

Ultimate mutations are in lower case (base number assignments are for wild-type sequence), flanking sequences in upper case, and the inverted repeat is doubly underlined. Although this mutation was recovered in other genetic backgrounds, it was most frequently encountered in a *mutB* mismatch repair defective strain. Such illegitimate synthesis could also explain the +TGT mutation at D872 and concomitant 5-bp deletion (AACTG):

D869<u>CGG</u>tgt<u>CCGCGCGCGGACACCG</u>

The role that quasi- or imperfect palindromes play in the generation of complex spontaneous mutations in other systems has previously been emphasized.[48–50] In the words of Jan Drake,[50] "when a palindrome is both imperfect and single-stranded, as in DNA replication, repair or recombination, it tends to form a stem bearing mismatches. If these mismatches are eliminated in any of several possible ways, mutations result." The two mutations shown in Figure 2.6B and D underscore how quasipalindromes may "direct" complex mutations in the *hisD3052* system and how illegitimate replication and/or repair may be among the "several possible ways"[50] that these mutations arise.

Clearly, of the four targets discussed, the *hisD3052* target is by far the most intricate. Without the use of specific probes to sweep away common reversion events, the intricacy of this target might not have been revealed.

ACKNOWLEDGMENTS

I thank Eugene Kupchella for assistance with Figures 2.1 and 2.2 and for sharing some unpublished data. I wish also to recognize Eugene Kupchella, Erik Henrikson, Dr. Michael J. Prival, and Dr. Walter H. Koch for their enthusiastic and productive endeavors in further developing the *Salmonella typhimurium his* paradigm. Finally, I wish to acknowledge Dr. Douglas L. Archer for his unwavering support of this research effort.

The author wishes to note that this manuscript has not been altered since the original submission in early 1993. Since that time, a number of studies from the author's laboratory, as well as others, have been published, reinforcing the usefulness of this approach.

REFERENCES

1. Ames, B. N., McCann, J., and Yamasaki, E. Methods for detecting carcinogens and mutagens with the Salmonella/mammalian-microsome mutagenicity test. *Mut. Res.* 31, 347, 1975.
2. Maron, D. M. and Ames, B. N. Revised methods for the *Salmonella* mutagenicity test. *Mut. Res.* 113, 173, 1983.
3. Prival, M. J. The *Salmonella* mutagenicity assay: Promises and problems. *Ann. N.Y. Acad. Sci.* 407, 154, 1983.
4. Hartman, P. E., Ames, B. N., Roth, J. R., Barnes, W. M., and Levin, D. E. Target sequences for mutagenesis in *Salmonella* histidine-requiring mutants. *Environ. Mutagen.* 8, 631, 1986.
5. Cebula, T. A. and Koch, W. H. Analysis of spontaneous and psoralen-induced *Salmonella typhimurium hisG46* revertants by oligodeoxyribonucleotide colony hybridization: Use of psoralens to cross-link probes to target sequences. *Mut. Res.* 229, 79, 1990.
6. Prival, M. J. and Cebula, T. A. Sequence analysis of mutations arising during prolonged starvation of *Salmonella typhimurium*. *Genetics* 132, 303, 1992.
7. Henrikson, E. N., Koch, W. H., and Cebula, T. A. Base substitution specificity of *hisG46* reversion in *S. typhimurium* strain TA100 uniquely characterizes several classes of mutagens. *Environ. Mol. Mutagen.* 19:S20, 24, 1992.
8. Cebula, T. A. and Koch, W. H. Sequence analysis of *Salmonella typhimurium* revertants. *Mutation and the Environment, Part D: Carcinogenesis*. Mendelsohn, M.L. and Albertini, R.J., Eds. Wiley-Liss, New York, 1990, p. 367.
9. Koch, W. H., Henrikson, E. N., and Cebula, T. A. Mutational specificity at the *hisC3076* locus of *Salmonella typhimurium*. *Environ. Mol. Mutagen.* 17:S19, 37, 1991.
10. Kupchella, E. and Cebula, T. A. Analysis of *Salmonella typhimurium hisD3052* revertants: The use of oligodeoxyribonucleotide colony hybridization, PCR, and direct sequencing in mutational analysis. *Environ. Mol. Mutagen.* 18, 224, 1991.
11. Cebula, T. A. and Koch, W. H. Polymerase chain reaction (PCR) and its application to mutational analysis. *New Horizons in Biological Dosimetry*. Gledhill, B. and Mauro, F., Eds. Wiley-Liss, New York, 1991, p. 255.
12. Wallace, R. B., Murphy, R. F., Bonner, J., and Itakura, K. Hybridization of synthetic oligodeoxyribonucleotides to $\Phi\chi$ 174 DNA: The effect of single base pair mismatch. *Nucleic Acids Res.* 6, 3543, 1979.
13. Wallace, R. B., Johnson, M. J., Hirose, T., Miyake, T., Kawashima, E. H., and Itakura, K. The use of synthetic oligonucleotides as hybridization probes. II. Hybridization of oligonucleotides of mixed sequence to rabbit β-globin DNA. *Nucleic Acids Res.* 9, 879, 1981.
14. Modrich, P. DNA mismatch correction. *Annu. Rev. Biochem.* 56, 435, 1987.
15. Werntges, H., Steger, G., Riesner, D., and Fritz, H.-J. Mismatches in DNA double strands: Thermodynamic parameters and their correlation to repair efficiencies. *Nucleic Acids Res.* 14, 3773, 1986.
16. Dalbadie-McFarland, G., Cohen, L. W., Riggs, A. D., Morin, C., Itakura, K., and Richards, J. H. Oligonucleotide-directed mutagenesis as a general and powerful method for studies of protein function. *Proc. Natl. Acad. Sci. U.S.A.* 79, 6409, 1982.
17. Wallace, R. B. and Miyada, C. G. Oligonucleotide probes for the screening of recombinant DNA libraries. *Methods Enzymol.* 152, 432, 1987.

18. Koch, W. H., Wentz, B. A., Trucksess, M. W., and Cebula, T. A. A nonisotopic DNA hybridization methodology for characterization of molecular alterations at *Salmonella his* loci. *Environ. Mol. Mutagen.* 17:S19, 37, 1991.
19. Maniatis, T., Fritsch, E. F., and Sambrook, J. *Molecular Cloning: A Laboratory Manual.* Cold Spring Harbor Laboratory, New York, 1982.
20. Sanger, F., Nicklen, S., and Coulsen, A. R. DNA sequencing with chain-terminating inhibitors. *Proc. Natl. Acad. Sci. U.S.A.* 74, 5463, 1977.
21. Saiki, R., Scharf, S., Faloona, F., Mullis, K. B., Horn, G. T., Erlich, H. A., and Arnheim, N. Enzymatic amplification of β-globin genomic sequences and restriction site analysis for diagnosis of sickle cell anemia. *Science* 230, 1350, 1985.
22. Gyllensten, U. B. and Erlich, H. A. Generation of single-stranded DNA by the polymerase chain reaction and its application to direct sequencing of the HLA DQA locus. *Proc. Natl. Acad. Sci. U.S.A.* 85, 7652, 1988.
23. Hartman, P. E., Hartman, Z., Stahl, R. C., and Ames, B. N. Classification and mapping of spontaneous and induced mutations in the histidine operon of *Salmonella. Adv. Genet.* 16, 1, 1971.
24. Barnes, W. M., Tuley, E., and Eisenstadt, E. Base-sequence analysis of His⁺ revertants of the *hisG46* missense mutation in *Salmonella typhimurium. Environ. Mutagen.* 4, 297, 1982.
25. Miller, J. K. and Barnes, W. M. Phenotypic and reversion analysis of a *Salmonella typhimurium* constructed to have an arginine codon at the *hisG46* missense codon. *Mut. Res.* 201, 189, 1988.
26. Levin, D. E. and Ames, B. N. Classifying mutagens as to their specificity in causing the six possible transitions and transversions: A simple analysis using the *Salmonella* mutagenicity assay. *Environ. Mutagen.* 8, 9, 1986.
27. Kupchella, E., Koch, W. H., and Cebula, T. A. Mutant alleles of tRNA^Thr genes are responsible for intergenic suppression of the *hisG46* missense mutation in *Salmonella typhimurium. Environ. Mol. Mutagen.* 23, 81, 1994.
28. Miller, J. K. and Barnes, W. M. Colony probing as an alternative to standard sequencing as a means of direct analysis of chromosomal DNA to determine the spectrum of single-base changes in regions of known sequence. *Proc. Natl. Acad. Sci. U.S.A.* 83, 1026, 1986.
29. Eisenstadt, E., Miller, J. K., Kahng, L.-S., and Barnes, W. M. Influence of *uvrB* and pKM101 on the spectrum of spontaneous, UV- and γ-ray-induced base substitutions that revert *hisG46* in *Salmonella typhimurium. Mut. Res.* 210, 113, 1990.
30. Wink, D. A., Kazimierz, K. S., Kasprzak, S., Maragos, C. M., Elespuru, R. K., Misra, M., Dunams, T. M., Cebula, T. A., Koch, W. H., Andrews, A. W., Allen, J. S., and Keefer, L. K. DNA deaminating ability and genotoxicity of nitric oxide and its progenitors. *Science* 254, 1001, 1991.
31. Hoppe, I., Johnston, H. M., Biek, D., and Roth, J. R. A refined map of the *hisG* gene of *Salmonella typhimurium. Genetics* 92, 17, 1979.
32. Levin, D. E., Hollstein, M., Christman, M. F., Schwiers, E. A., and Ames, B. N. A new *Salmonella* tester strain (TA102) with A·T base pairs at the site of the mutation detects oxidative mutagens. *Proc. Natl. Acad. Sci. U.S.A.* 79, 7445, 1982.
33. Levin, D. E., Marnett, L. J., and Ames, B. N. Spontaneous and mutagen-induced deletions; mechanistic studies in *Salmonella* tester strain TA102. *Proc. Natl. Acad. Sci. U.S.A.* 81, 4457, 1984.

34. Guttenplan, J. B., Henrikson, E. N., and Cebula, T. A. Effects of 2-hydroxylation on the mutagenic activity and specificity of alkyldiazonium ions: mechanistic deductions from mutational data. *Environ. Mol. Mutagen.* 19:S20, 22, 1992.

35. Koch, W. H. and Cebula, T. A. pKM101 induces complex frameshift events in the *Salmonella typhimurium hisC3076* locus. *Environ. Mol. Mutagen.* 15:S17, 30, 1990.

36. Oeschger, N. S. and Hartman, P. E. ICR-induced frameshift mutations in the histidine operon of *Salmonella J. Bacteriol.* 101, 490, 1970.

37. Isono, K. and Yourno, J. Chemical carcinogens as frameshift mutagens: *Salmonella* DNA sequence sensitive to mutagenesis by polycyclic carcinogens. *Proc. Natl. Acad. Sci. U.S.A.* 71, 1612, 1974.

38. Cebula, T. A., Payne, W. L., Trucksess, M. W., and Hill, W. E. Colony probing of *Salmonella typhimurium hisD3052* revertants. *Environ. Mol. Mutagen.* 9:S8, 23, 1987.

39. Fuscoe, J. C., Wu, R., Shen, N. H., Healy, S. K., and Felton, J. S. Base-change analysis of revertants of the *hisD3052* allele in *Salmonella typhimurium. Mut. Res.* 201, 241, 1987.

40. Bell, D. A., Levine, J. G., and DeMarini, D. M. DNA sequence analysis of revertants of the *hisD3052* allele of *Salmonella typhimurium* TA98 using the polymerase chain reaction and direct sequencing: Application to 1-nitropyrene-induced revertants. *Mut. Res.* 252, 35, 1991.

41. O'Hara, S. M. and Marnett, L. J. DNA sequence analysis of spontaneous and β-methoxy-acrolein-induced mutations in *Salmonella typhimurium hisD3052. Mut. Res.* 247, 45, 1991.

42. DeMarini, D. M., Abu-Shakra, A., Gupta, R., Hendee, L. J., and Levine, J. G. Molecular analysis of mutations induced by the intercalating agent ellipticine at the *hisD3052* allele of *Salmonella typhimurium* TA98. *Environ. Mol. Mutagen.* 20, 12, 1992.

43. Streisinger, G., Okada, Y., Emrich, J., Newton, J., Tsugita, A., Terzaghi, E., and Inouye, M. Frameshift mutations and the genetic code. *Cold Spring Harbor Symp. Quant. Biol.* 31, 77, 1966.

44. Kunkel, T. A. and Soni, A. Mutagenesis by transient misalignment. *J. Biol. Chem.* 263, 14784, 1988.

45. Bebenek, K. and Kunkel, T. A. Frameshift errors initiated by nucleotide misincorporation. *Proc. Natl. Acad. Sci. U.S.A.* 83, 4946, 1990.

46. Farabaugh, P. J., Schmeissner, U., Hofer, M., and Miller, J. H. Genetic studies of the *lac* repressor. VII. On the molecular nature of spontaneous hotspots in the *lacI* gene of *Escherichia coli. J. Mol. Biol.* 126, 847, 1978.

47. Kupchella, E. and Cebula, T. A. Mitomycin C induces complex mutations in *Salmonella* strain TA1978/pKM101. *Environ. Mol. Mutagen.* 15:S17, 33, 1990.

48. Ripley, L. S. Frameshift mutation: Determinants of specificity. *Annu. Rev. Genet.* 24, 189, 1990.

49. Todd, P. A. and Glickman, B. W. Mutational specificity of UV light in *Escherichia coli*: Indications for a role of DNA secondary structure. *Proc. Natl. Acad. Sci. U.S.A.* 79, 4123, 1982.

50. Drake, J. W. Spontaneous mutation. *Annu. Rev. Genet.* 25, 125, 1991.

CHAPTER 3

Mutational Spectra of the Four Bay-Region Diol Epoxides of Benzo[c]phenanthrene and the Drinking Water Mutagen, MX, Determined with an Improved Set of *E. coli lacZ⁻* Mutants

Chuang Lu, Haruhiko Yagi, Donald M. Jerina, and Albert M. Cheh

INTRODUCTION

Polycyclic aromatic hydrocarbons (PAH) are common environmental carcinogens.[1] They are metabolically transformed to bay-region diol epoxides that act as ultimate carcinogens that covalently bind to DNA[2] by N-alkylation of the exocyclic amino groups of the purine bases. The resulting bound adducts are thought to initiate carcinogenesis and mutagenesis. Four optically active diol epoxides (enantiomers of a pair of diastereomers in which the benzylic hydroxyl group and epoxide oxygen are either *cis*, DE1, or *trans*, DE2, Fig. 3.1) may be formed from a given hydrocarbon. In general, only selected isomers of the bay-region diol epoxides from a given hydrocarbon [typically the enantiomer with (R,S,S,R)-absolute configuration reading from the benzylic hydroxyl carbon to the benzylic epoxide carbon] possess high tumorigenic activity.[3] The four bay-region diol epoxide isomers of benzo[a]pyrene (BP),[4] benzo[c]phenanthrene (BcPh, Fig. 3.1),[5] dibenz[a,j]anthracene,[6] and benz[a]anthracene[7] all generate different patterns of DNA adducts. Major differences occur in the ratio of deoxyguanosine to deoxyadenosine adducts and the extent of *cis* vs. *trans* opening of the epoxide ring at the benzylic position; however, there has been only limited success in drawing correlations between the formation of specific DNA adducts and the carcinogenicity of the diol epoxide isomer.[7] On the other hand, a correlation between mutation (resulting from adduct formation) and carcinogenesis is suggested by, e.g., the observation that a point mutation is responsible for the activation of *ras* oncogenes in PAH-induced tumors.[8] Six

0-87371-951-4/95/$0.00+$.50
© 1995 by CRC Press, Inc.

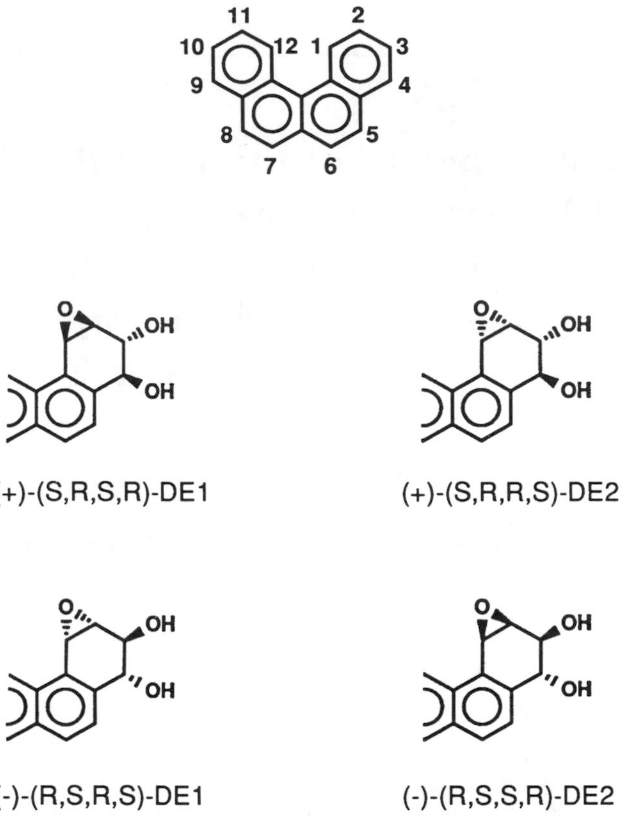

(+)-(S,R,S,R)-DE1   (+)-(S,R,R,S)-DE2

(-)-(R,S,R,S)-DE1   (-)-(R,S,S,R)-DE2

Figure 3.1   Numbering for benzo[c]phenanthrene (BcPh), and the structures of its optically active
           bay-region 3,4-diol 1,2-epoxides. Absolute configurations are designated starting from
           the benzylic hydroxyl carbon and progressing to the benzylic epoxide carbon.

different base substitution mutations are possible: G to C, A, or T; and A to C,
G, or T. The relative amounts of each that are formed by a compound comprise
its mutational spectrum. It is desirable to determine the mutational spectra of
individual PAH diol epoxide isomers to correlate the patterns of adduct formation and relative carcinogenicities exhibited by the respective isomers.

Mutational spectra are often determined by creating mutations within a
specific gene, identifying the mutants via a selection process, amplifying the
mutant DNA, sequencing to determine the base change, and tabulating the
types of base changes that are observed. Several mutational spectra determined
mostly in this manner have been reported for racemic mixtures of diol epoxide
isomers [usually (±)-BP DE2],[9-12] but comparatively few mutational spectra
have been determined with optically pure isomers. Published mutational spectra exist for all four optically pure 3,4-diol 1,2-epoxide isomers of BcPh[13] and the
most carcinogenic (+)-(7R,8S)-diol (9S,10R)-epoxide isomer of BP.[14]

The labor-intensive nature of determining mutational spectra via this process has led to the development of more rapid techniques based on characterizing the reverse mutations that occur at single sites such as the *hisG46* site

Table 3.1   Bacterial Strains, Base Substitutions, and Mutagens That Revert Them

Cupples and Miller strain[a]	Improved strain[b]	Base substitution	Diagnostic mutagen[c]
CC101	CL101P	AT → CG	Angelicin + UV light
CC102	CL102P	GC → AT	Ethylmethanesulfonate
CC103	CL103P	GC → CG	5-Azacytidine
CC104	CL104P	GC → TA	Methylglyoxal
CC105	CL105P	AT → TA	t-Butyl hydroperoxide
CC106	CL106P	AT → GC	N^4-Hydroxycytidine

[a] All the CC strains[16] are *E. coli* that are Δ(*lac proB*) *ara⁻ thi⁻* and contain an episome that is *lacI⁻ Z⁻ proB⁺*; CC101 is also *nal^r*.

[b] The CLP strains are *E. coli* that are Δ(*lac proB*) Δ(*uvrB-bio*) *ara⁻ thi⁻* and contain the *lacI⁻ Z⁻ proB⁺* episome, pKM101, and an unknown mutation (*rfa?*) that increases bacterial permeability to high molecular weight compounds;[9] CL101P is also *nal^r*.

[c] From Levin and Ames.[15]

present in several Ames *Salmonella typhimurium* tester strains. Such an approach trades the ability to identify the sequence specificity of the mutations (when they are identified by sequencing methodologies), for far greater speed in scoring the types of mutations when they are produced at only one or a few known codons within a sequence. Only mutants grow to form colonies on an agar plate, thus simplifying their identification. Levin and Ames[15] developed a set of phenotypic screens for mutations produced in a set of *S. typhimurium* strains; the screens could be used to distinguish in a semiquantitative fashion the principal types of mutations induced by a compound. Cebula and Koch[16] hybridized mutant sequences with deoxynucleotide oligomers in a rapid and simple procedure for identifying and tabulating sequence changes in *S. typhimurium hisG46* revertants.

Cupples and Miller[17] constructed a series of six *Escherichia coli lacZ⁻* strains, CC101–CC106, each of which is reverted to Lac⁺ prototrophy by one of the six possible base pair substitutions. This promised to make the determination of mutational spectra very simple. The six strains would be treated with a mutagen in parallel, and by scoring the number of revertants generated with each strain, the mutational spectrum would be determined. Table 3.1 shows the six *E. coli* strains developed by Cupples and Miller,[17] the base pair substitution that reverts each of the strains,[17] and the diagnostic mutagen that was reported by Levin and Ames[15] to generate specifically that type of mutation. It was found that the original *E. coli* CC101–CC106 strains failed to respond to several of the diagnostic mutagens until their *lacZ⁻* containing episomes were placed in a genetic background of increased permeability to mutagens (an *rfa*-like mutation), Δ*uvrB* (to eliminate excision repair), and pKM101 (to enhance the SOS-dependent mutagenic response) to create an improved tester set called CL101P–CL106P (Table 3.1).[18] Ames et al.[19,20] had previously done exactly this to enhance the response of their *S. typhimurium* tester strains to mutagens. This paper reports the mutational spectra of the four 3,4-diol 1,2-epoxide isomers of BcPh that were determined with the improved tester set of *E. coli lac Z⁻* strains.

The compound MX [3-chloro-4-(dichloromethyl)-5-hydroxy-2(5H)-furanone], a potent mutagen in the Ames *S. typhimurium* bacteria, was first identified as a by-product present in pulp chlorination effluent,[21] and later was found in chlorinated drinking waters. Various studies of its biological activity and its occurrence in drinking water have been reviewed by Meier;[22] it appears that MX could account for about one third of the overall mutagenic activity of chlorinated drinking waters in the U.S. Little is known about the types of DNA adducts or mutations that might be created by MX. The present study reports a preliminary evaluation of its mutational spectrum determined with the improved set of E. coli lac Z− strains.

MATERIALS AND METHODS

Creation of the E. coli CL101P–CL106P strains by transfer of the lacZ−-containing episomes of E. coli strains CC101–CC106[17] to E. coli strain EE122, and the responses of CL101P–CL106P to the set of mutagens specific for each type of base pair substitution is described elsewhere.[18] The CL101P–CL106P strains were grown at 37°C in LB broth[23] (10 g/L Bacto tryptone, 5 g/L Bacto yeast extract, and 10 g/L NaCl) with shaking.

Mutagenicities of test compounds were determined with a modified Ames test preincubation procedure[24] as follows: Aliquots of overnight LB broth cultures of bacteria (0.1 mL) were mixed with pH 7.0 minimum A buffer[23] [0.5 mL of 10.5 g/L K_2HPO_4, 4.5 g/L KH_2PO_4, 1 g/L $(NH_4)_2SO_4$, and 0.5 g/L sodium citrate · $2H_2O$] and the polycyclic aromatic hydrocarbon diol epoxide (in not more than 30 µL of acetonitrile, and coded for blind testing) for approximately 30 min at 37°C. When MX was tested, it was added in not more than 30 µL of ethanol instead of acetonitrile. Because of its instability at nonacidic pH,[22] 0.5 mL of 0.1 M sodium citrate buffer, pH 5.5, was used in place of the 0.5 mL volume of pH 7.0 minimum A buffer. Then 2 mL of top agar (8 g/L agar, 8 g/L NaCl, and 15 mg/L biotin) was added, and the mixture was poured onto a minimum A plate (15 g/L agar, 5 mg/L thiamine, 0.2 g/L $MgSO_4$ · $7H_2O$, and 2 g/L D-lactose added to minimum A buffer). Revertant colonies were counted after 3 days at 37°C; if 250 mg/L of glucose was added to the top agar (analogous to the addition of a trace of histidine to the top agar in the Ames test),[24] plates could be counted after 2 days. To construct mutational spectra, all six strains were simultaneously treated and plated in an identical fashion. Individual dose–response experiments were performed by plating three doses and a solvent control, each in triplicate, and calculating the results (in revertants per microgram of diol epoxide or nanogram of MX) with a linear regression fit of the data. To determine the mutational spectra of the four BcPh diol epoxide isomers, the plate counts from three replicate experiments were combined prior to the regression analysis.

RESULTS AND DISCUSSION

The mutagenesis procedure utilized by Cupples and Miller[17] with the original six CC101–CC106 strains involved treating a washed log phase culture with

Table 3.2 Reversion of the Improved Set of *E. coli lacZ*⁻ Mutants by Four BcPh Diol Epoxide Isomers

Isomer/dose	Revertants per Plate[a]					
	CL101P AT → CG	CL102P GC → AT	CL103P GC → CG	CL104P GC → TA	CL105P AT→ TA	CL106P AT → GC
Solvent control	23 ± 9	34 ± 10	6 ± 8	99 ± 25	30 ± 17	6 ± 5
(−)-(R,S,R,S)-DE1						
0.08 µg	73 ± 23	70 ± 17	11 ± 8	523 ± 167	206 ± 35	38 ± 10
0.16 µg	92 ± 10	83 ± 18	15 ± 5	601 ± 170	307 ± 51	39 ± 7
0.24 µg	123 ± 20	106 ± 34	21 ± 7	661 ± 195	308 ± 72	50 ± 17
(+)-(S,R,S,R)-DE1						
0.05 µg	51 ± 8	54 ± 13	11 ± 9	188 ± 29	209 ± 58	15 ± 5
0.10 µg	83 ± 11	65 ± 13	14 ± 10	225 ± 45	270 ± 74	19 ± 5
0.15 µg	96 ± 12	73 ± 34	14 ± 6	236 ± 45	285 ± 78	27 ± 12
(+)-(S,R,R,S)-DE2						
0.05 µg	74 ± 20	222 ± 98	11 ± 11	511 ± 111	255 ± 124	30 ± 15
0.10 µg	99 ± 34	288 ± 116	14 ± 10	633 ± 191	310 ± 129	25 ± 12
0.15 µg	139 ± 49	290 ± 110	15 ± 10	646 ± 150	328 ± 132	34 ± 15
(−)-(R,S,S,R)-DE2						
0.05 µg	65 ± 26	114 ± 31	14 ± 12	231 ± 60	725 ± 155	19 ± 6
0.10 µg	87 ± 28	123 ± 41	14 ± 9	229 ± 59	735 ± 224	19 ± 7
0.15 mg	117 ± 26	120 ± 23	14 ± 6	214 ± 54	720 ± 188	26 ± 10

[a] Revertants per plate show a composite mean ±standard deviation for each dose and bacterial strain, obtained from three separate experiments with three replicate plates each for a total of nine plates.

mutagen, washing with buffer to remove unreacted mutagen, determining the percent of bacteria surviving, growing them in rich medium to express mutants, and then plating to determine numbers of mutants per 10^8 survivors. A simpler procedure is used in the Ames test, in which a mutagen is added to a stationary overnight culture of bacteria plus top agar, and the mixture is poured onto plates directly. To construct a mutational spectrum, the six *E. coli* strains must be treated in parallel; with the simpler procedure it is easier for a single individual to use multiple doses and perform replicates of each dose to observe a dose–response (at the expense of being unable to determine mutant frequencies per number of survivors, because the immediate plating of the reaction mixture makes it impossible to determine the number of survivors); therefore, the simpler procedure was utilized in this study.

Wood et al.[25] reported that preincubation was necessary to obtain the maximum Ames test response to PAH diol epoxides; in the preincubation method that was applied (Materials and Methods), the concentration of organic solvent was kept below 10%, an amount that was neither toxic nor affected mutagenesis. Table 3.2 lists the responses of the improved *E. coli* strains CL101P–CL106P to the four diol epoxide isomers of BcPh; the data therein are combined from three replicate experiments. The doses used were determined by the responses in a preliminary dose ranging study whose results are not included in Table 3.2 because different doses were used and fewer replicates were done.

Table 3.3 Relative Amounts of Base Substitution Mutations Produced by Four BcPh Diol Epoxide Isomers: Comparison of *E. coli lacZ⁻* and *supF* Data[a]

Diol epoxide isomer	CL101P AT → CG	CL102P GC → AT	CL103P GC → CG	CL104P GC → TA	CL105P AT → TA	CL106P AT → GC
(−)-(R,S,R,S)-DE1						
lacZ⁻ revertants/µg	398	282	59	2215	1170	168
lacZ⁻ (%)	9.3	6.6	1.4	51.7	27.3	3.9
supF (%)	6	7	18	38	27	3
(+)-(S,R,S,R)-DE1						
lacZ⁻ revertants/µg	501	254	49	906	1655	135
lacZ⁻ (%)	14.3	7.3	1.4	25.9	47.3	3.9
supF (%)	10	8	10	14	50	8
(+)-(S,R,R,S)-DE2						
lacZ⁻ revertants/µg	744	1665	56	3537	1899	158
lacZ⁻ (%)	9.2	20.7	0.7	43.9	23.6	2.0
supF (%)	4	19	16	39	20	2
(−)-(R,S,S,R)-DE2						
lacZ⁻ revertants/µg	604	529	46	702	4196	118
lacZ⁻ (%)	9.7	8.5	0.7	11.3	67.7	1.9
supF (%)	7	4	16	33	31	9

[a] Revertants/µg of diol epoxide represent the slopes of dose–response curves, calculated by linear regression from the plate counts also used to construct Table 3.2. The percent of each type of base substitution (*E. coli lacZ⁻*) is then compared with the percent reported in Bigger et al.[13] (*supF*).

Table 3.2 shows that the individual BcPh diol epoxide isomers revert the six *lacZ⁻* strains CL101P–CL106P to different degrees. The standard deviations are fairly sizable percentages of the mean revertants per plates because of the often sizable day-to-day variation in the number of revertants seen on both the solvent control and diol epoxide-treated plates. However, from one day to another, the plate counts, whether they were from solvent control plates or mutagen-treated ones, generally increased or decreased in tandem across all six strains, so the revertants per microgram diol epoxide and the relative responses in each strain compared to the others (which is the basis for determining the mutational spectrum) changed less than the standard deviations would indicate.

Table 3.3 compares the mutational spectra of four BcPh diol epoxide isomers obtained with the *E. coli lac Z⁻* strains, CL101P–CL106P, to the published mutational spectra obtained with a *supF* shuttle vector target.[13] For each diol epoxide isomer, the revertants per microgram of diol epoxide observed in each strain of the *lacZ⁻* system are shown first; these are based on linear regression analyses of the plate counts, which are summarized in Table 3.2. Below these are shown the percent of each type of mutation that was observed with the *lacZ⁻* and *supF* systems, respectively. The major difference between the the responses of the two systems is the very low numbers of GC → CG mutations scored in the *lacZ⁻* system; the percents are about an order of magnitude lower than in the *supF* system. Other than this, there is fairly close agreement between the two systems. For each of the diol epoxide isomers, the rank order of each type of base substitution other than GC → CG is virtually identical in *lacZ⁻* compared to *supF*.

In the *lacZ*⁻ system the most carcinogenic BcPh diol epoxide isomer, (–)-(*R,S,S,R*)-DE2, produced predominately (about two thirds) AT → TA transversions, followed by about 10% each of GC → TA and AT → CG transversions and GC → AT transitions. In the *supF* system the same isomer produced almost equal amounts of AT → TA (31%) and GC → TA (33%) transversions, 9% AT → GC transitions, 7% AT → CG transversions, and only 4% GC → AT transitions. With the other diol epoxide isomers, except for GC → CG transversions, similar amounts of each of the base substitutions were observed in the *lacZ*⁻ vs. *supF* systems. The (+)-(*S,R,R,S*)-DE2 isomer generated about 40% GC → TA transversions, 20% AT → TA transversions or GC → AT transitions, and lesser amounts of AT → CG transversions and AT → GC transitions in both systems. The (–)-(*R,S,R,S*)-DE1 isomer generated about 40–50% GC → TA and 27% AT → TA transversions, 6–9% AT → CG transversions and GC → AT transitions, and 3–4% AT → GC transitions in both systems. The virtually noncarcinogenic (+)-(*S,R,S,R*)-DE1 isomer formed about 50% AT → TA and 14–26% GC → TA transversions, and 4–15% AT → CG transversions and GC → AT and AT → GC transitions in the two systems.

It can be seen in Table 3.2 that the dose–response in the *E. coli lacZ*⁻ system flattened with higher doses. It is not known why this occurred, since a plate count that reached a plateau with one diol epoxide isomer reached a much higher plateau with a different one, and much higher plate counts have been observed with other types of mutagens (see the MX data below), arguing against a flattening due to toxicity at higher doses. A possible explanation could be the limited solubility of the diol epoxides in aqueous solution. In any event, computing the dose–response with the solvent control and only the lowest dose of diol epoxide instead of all the doses does not change the mutational spectra observed with the *E. coli lacZ*⁻ strains to an appreciable degree. The only change in rank order of the kinds of base substitutions observed with the four diol epoxides is an increase in the GC → AT transition percent observed with (–)-(*R,S,S,R*)-DE2, to put it ahead of the amount of AT → CG transversions.

In summary, the BcPh diol epoxide mutational spectra observed with the improved *E. coli lacZ*⁻ strains correlates well with those reported with the *supF* shuttle vector target, with the exception that the response of strain CL103P (GC → CG transversions) is lower by an order of magnitude. The negative control reversion rate of this strain was considerably lower than the rate for most of the other strains, suggesting it might be a cold spot for mutation. The control rate with CL106P was also low, but AT → GC transitions were not observed in high frequency with BcPh diol epoxides in the *supF* system, and the *lacZ*⁻ percents of AT → GC transitions appear to be much more in line with the *supF* percents, so any speculation about CL106P also being cold is currently unwarranted.

It has been noted[26] that the DNA adducts formed by PAH diol epoxides follow the "A rule," whereby deoxyadenosine is inserted by DNA polymerase across from bulky adducts. This can be seen in the *supF* data[13] (Table 3.3, which show GC → TA and AT → TA transversions to be 60–65% of the base substitutions) and in other PAH diol epoxide mutational spectra.[9–12,14] The same is seen with the *lacZ*⁻ system, where it would appear that an even higher propor-

Table 3.4 Reversion of the Improved Set of *E. coli lacZ*- Mutants by MX

Dose	Revertants per plate[a]					
	CL101P AT → CG	CL102P GC → AT	CL103P GC → CG	CL104P GC → TA	CL105P AT → TA	CL106P AT → GC
Solvent control	14 ± 1	50 ± 2	0 ± 0	117 ± 7	40 ± 6	2 ± 2
5 ng	106 ± 6	52 ± 7	1 ± 0	364 ± 23	43 ± 1	4 ± 1
15 ng	491 ± 50	85 ± 5	15 ± 2	1064 ± 43	54 ± 9	5 ± 2
30 ng	1271 ± 41	116 ± 6	40 ± 4	2043 ± 24	101 ± 19	5 ± 1
Revertants/ng[b]	43	2.3	1.4	65	2.1	<0.1
SE[c]	2.1	0.2	0.1	1.1	0.4	—
Percent[d]	38	2.0	1.2	57	1.8	0

[a] Means ± standard deviations are shown for three replicate plates.

[b] Calculated by linear regression from the plate counts for the solvent control plates and the plates receiving the three doses.

[c] Standard error of the revertants/ng.

[d] (Percent) of total base substitutions that correspond to this type.

tion of transversions to TA is formed than in the *supF* system (Table 3.3), except such a conclusion must be tempered by uncertainty as to the true proportion of GC → CG transversions.

A preliminary experiment was performed with the improved *E. coli lacZ*-strains to determine the mutational spectrum of MX. The results are shown in Table 3.4. Unlike the BcPh diol epoxides, MX gave a linear response with the doses tested (except with strain CL106P, which did not respond), with some plate counts in the thousands. GC → TA transversions were the most abundant base substitutions (57%), followed by large numbers of AT → CG transversions (38%) and very few of the other types. It is not known if the number of GC → CG transversions is underestimated because of poor responses by CL103P. Regardless of this, the far greater number of AT → CG compared to AT → TA mutations observed suggests that unlike the PAH diol epoxides, MX does not follow the "A rule" with adducts that are presumed to have been formed at A.

ACKNOWLEDGMENTS

We thank Drs. Jeffrey H. Miller, Michael J. Prival, and Thomas A. Cebula for their gifts of *E. coli* strains. This work was supported in part by NIH Grant 1R15 CA52043 to A.M.C.

REFERENCES

1. Dipple, A., Moschel, R. C., and Bigger, C. A. H. Polycyclic aromatic hydrocarbons. *Chemical Carcinogens.* Searle, C. E., Ed. ACS Monograph 182. American Chemical Society, Washington, D.C., 1984, Vol. 1, p. 41.

2. Jerina, D. M. and Lehr, R. E. The bay-region theory: A quantum mechanical approach to aromatic hydrocarbon-induced carcinogenicity. *Microsomes and Drug Oxidations: 3rd International Symposium.* Ullrich, V., Roots, I., Hildebrandt, A. G., Estabrook, R. W., and Conney, A. H., Eds. Pergamon Press, Oxford, England, 1977, p. 709.

3. Jerina, D. M., Sayer, J. M., Agarwal, S. K., Yagi, H., Levin, W., Wood, A. W., Conney, A. H., Pruess-Schwartz, D., Baird, W. M., Pigott, M. A., and Dipple, A. Reactivity and tumorigenicity of bay-region diol epoxides derived from polycyclic aromatic hydrocarbons. *Biological Reactive Intermediates III.* Kocsis, J. J., Jollow, D. J., Witmer, C. M., Nelson, J. O., and Snyder, R., Eds. Plenum Press, New York, 1986, p. 11.

4. Sayer, J. M., Chadha, A., Agarwal, S. K., Yeh, H. J. C., Yagi, H., and Jerina, D. M. Covalent nucleoside adducts of benzo[a]pyrene 7,8-diol 9,10-epoxides: Structural reinvestigation and characterization of a novel adenosine adduct on the ribose moiety, *J. Org. Chem.* 56, 20, 1991.

5. Dipple, A., Pigott, M. A., Agarwal, S. K., Yagi, H., Sayer, J. M., and Jerina, D. M. Optically active benzo[c]phenanthrene diol epoxides bind extensively to adenine residues in DNA. *Nature (London)* 327, 535, 1987.

6. Chadha, A., Sayer, J. M., Yeh, H. J. C., Yagi, H., Cheh, A. M., Pannell, L. K., and Jerina, D. M. Structures of covalent nucleoside adducts formed from adenine, guanine and cytosine bases of DNA and the optically active bay-region 3,4-diol 1,2-epoxides of dibenz[a,j]anthracene. *J. Am. Chem. Soc.* 111, 5456, 1989.

7. Cheh, A. M., Chadha, A., Sayer, J. M., Yeh, H. J. C., Yagi, H., Pannell, L. K., and Jerina, D. M. Structures of covalent nucleoside adducts formed from adenine, guanine and cytosine bases of DNA and the optically active bay-region 3,4-diol 1,2-epoxides of benz[a]anthracene. *J. Org. Chem.* 58, 4013, 1993.

8. Barbacid, M. Oncogenes and human cancer: cause or consequence? *Carcinogenesis* 7, 1037, 1986.

9. Eisenstadt, E., Warren, A. J., Porter, J., Atkins, D., and Miller, J. H. Carcinogenic epoxides of benzo[a]pyrene and cyclopenta[cd]pyrene induce base substitutions via specific transversions. *Proc. Natl. Acad. Sci. U.S.A.* 79, 1945, 1982.

10. Yang, J.-L., Maher, V. M., and McCormick, J. J., Kinds of mutations formed when a shuttle vector containing addducts of (±)-7β,8α-dihydroxy-9α,10α-epoxy-7,8,9,10-tetrahydrobenzo[a]pyrene replicates in human cells. *Proc. Natl. Acad. Sci. U.S.A.* 84, 3787, 1987.

11. Bernelot-Moens, C., Glickman, B. W., and Gordon, A. J. Induction of specific frameshift and base substitution events by benzo[a]pyrene diol epoxide in excision-repair-deficient *Escherichia coli. Carcinogenesis* 11, 781, 1990.

12. Yang, J.-L., Chen, R.-H., Maher, V. M., and McCormick, J. J. Kinds and location of mutations induced by (±)-7β,8α-dihydroxy-9α,10α-epoxy-7,8,9,10-tetrahydrobenzo[a]pyrene in the coding region of the hypoxanthine (guanine) phosphoribosyltransferase gene in diploid human fibroblasts. *Carcinogenesis* 12, 71, 1991.

13. Bigger, C. A. H., St. John, J., Yagi, H., Jerina, D. M., and Dipple, A. Mutagenic specificities of four stereoisomeric benzo[c]phenanthrene dihydrodiol epoxides. *Proc. Natl. Acad. Sci. U.S.A.* 89, 368, 1992.

14. Wei, S.-J. C., Chang, R. L., Wong, C.-Q., Bhachech, N., Cui, X. X., Hennig, E., Yagi, H., Sayer, J. M., Jerina, D. M., Preston, B. D., and Conney, A. H. Dose-dependent differences in the profile of mutations induced by an ultimate carcinogen from benzo[a]pyrene. *Proc. Natl. Acad. Sci. U.S.A.* 88, 11227, 1991.

15. Levin, D. E. and Ames, B. N. Classifying mutagens as to their specificity in causing the six possible transitions and transversions: A simple analysis using the *Salmonella* mutagenicity assay. *Environ. Mutagen.* 8, 9, 1986.

16. Cebula, T. A. and Koch, W. H. Analysis of spontaneous and psoralen-induced *Salmonella typhimurium hisG46* revertants by oligodeoxyribonucleotide colony hybridization: Use of psoralens to cross-link probes to target sequences. *Mut. Res.* 229, 79, 1990.

17. Cupples, C. G. and Miller, J. H. A set of *lacZ* mutations in *Escherichia coli* that allow rapid detection of each of the six base substitutions. *Proc. Natl. Acad. Sci. U.S.A.* 86, 5345, 1989.

18. Lu, C., Ogwuru, N., and Cheh, A. M. Manuscript in preparation.

19. Ames, B. N., Lee, F. D., and Durston, W. E. An improved bacterial test system for detection and classification of mutagens and carcinogens. *Proc. Natl. Acad. Sci. U.S.A.* 70, 782, 1973.

20. McCann, J., Spingarn, N. E., Kobori, J., and Ames, B. N. Detection of carcinogens as mutagens: Bacterial tester strains with R factor plasmids. *Proc. Natl. Acad. Sci. U.S.A.* 72, 979, 1975.

21. Holmbom, B. R., Voss, R. H., Mortimer, R. D., and Wong, A. Isolation and identification of an Ames-mutagenic compound present in kraft chlorination effluents. *Tappi* 64, 172, 1981.

22. Meier, J. R. Genotoxic activity of organic chemicals in drinking water. *Mut. Res.* 196, 211, 1988.

23. Miller, J. H. *Experiments in Molecular Genetics.* Cold Spring Harbor Laboratory, Cold Spring Harbor, NY, 1972.

24. Maron, D. M. and Ames, B. N., Revised methods for the *Salmonella* mutagenicity test. *Mut. Res.* 113, 173, 1983.

25. Wood, A. W., Chang, R. L., Levin, W., Ryan, D. E., Thomas, P. E., Croisy-Delcy, M., Ittah, Y., Yagi, H., Jerina, D. M., and Conney, A. H. Mutagenicity of the dihydrodiols and bay-region diol-epoxides of benzo[c]phenanthrene in bacterial and mammalian cells. *Cancer Res.* 40, 2876, 1980.

26. Strauss, B. S. The origin of point mutations in human tumor cells. *Cancer Res.* 52, 249, 1992.

CHAPTER 4

Assessment of Gene Expression in the Environment: Quantitative mRNA Analysis in Contaminated Soils

James T. Fleming and Gary S. Sayler

INTRODUCTION

The measurement of microbial activity and gene expression in the environment has posed a considerable challenge to microbiologists and microbial ecologists, where traditional pure mono-culture-based techniques commonly used in the laboratory are not suitable for the study of heterogeneous populations and organisms that are not readily culturable.[1,2] This is especially the case in complex matrixes such as soils and sediments, where interest in the bioremediation of hazardous wastes and concern surrounding the release of genetically engineered organisms have focused attention on *in situ* activity studies.[3,4] Bioremediation, where applicable as a waste treatment strategy, has the advantages of environmental friendliness and cost over conventional treatment regimens such as air stripping and incineration. The general acceptance of bioremediation has, however, been hindered by its unpredictable performance at the field-scale.[5] Companies in the business of hazardous waste remediation that are comfortable using mechanical or engineering approach processes such as incineration or sequestration/storage regard this new biotechnology as risky. This unpredictability of microbial breakdown of toxic compounds is primarily a biological problem rather than an engineering one; it merely reflects the present lack of fundamental knowledge about microbial gene expression and activity in natural environments. Before the biological processes involved may be engineered to increase *in situ* degradative rates, for example, by the addition of nutrients to or the aeration of contaminated soils, methods to accurately monitor the specific microbial degradative activities of interest must be developed.[6] This section will present a brief overview of traditional approaches to measuring microbial activities in natural environments with specific attention to soils and sediments. This will be followed by a description of *in situ* transcriptional analysis as a novel method to study microbial gene expression in soils.

TRADITIONAL APPROACHES FOR ASSESSING MICROBIAL ACTIVITY IN SOILS

Perhaps the most acceptable method, from an engineering standpoint, for following biodegradation of organic compounds in the environment has been a mass-balance determination of the transformation or mineralization of the target compounds based on treatability studies. Biotransformation rates may be derived from the analytical determination of initial and final pollutant concentrations from a laboratory scale study. However, the results of such an experiment are certainly not clear cut. Chemical analysis in complex nonhomogeneous matrices such as contaminated soils is difficult; analytical means may vary 100–200% of the mean of the replicates.[7] To further complicate the issue, competing abiotic mechanisms such as volatilization, sorption, or chemical reactions may be responsible for compound degradation.[7] As will affect all laboratory scale experiments mentioned in this chapter, extrapolation of mass-balance data from laboratory scale to field scale is complicated by the fact that compound fates and kinetics in the field may differ significantly from bench scale studies.

Based on wastewater treatment experience, the determination of biochemical oxygen demand (BOD) or chemical oxygen demand (COD) has been used to estimate microbial activity of polluted waters, soils, and sediments. This test measures the oxygen required for the biochemical degradation of organic materials and the oxidation of inorganic material such as sulfides and ferrous iron.[8] BOD has proved useful as a biodegradation parameter because it correlates well with bulk removal of pollutants. The shortcomings of this procedure are (1) it is not a specific indicator of organic oxidation, and (2) it yields no compound-specific degradation information. In addition, many organic chemicals of environmental importance are degraded by anaerobic mechanisms that are invisible to BOD testing.

The adenosine triphosphate (ATP) assay is a frequently used procedure in environmental microbiology that has also been applied to soil systems.[9] The determination of total intracellular ATP is a good biomass indicator and permits discrimination between viable and nonviable cells. In addition, ATP quantitation permits the physiological state of a microbial population to be determined using adenylate energy charge measurements. The main drawback of this procedure is that it is nonspecific in that it does not provide information about the identity of cells but only that they are active.[10]

In terms of relating degradation information to the biological processes involved the BOD, chemical analysis mass-balance and ATP approaches are relatively nonspecific; these techniques give no specific information about the particular organisms or biochemical pathways involved. In contrast to the nonspecificity of BOD and disappearance studies, determination of intermediary metabolites or radiolabeled mineralization studies offers a highly specific estimate of biodegradative activity in complex matrices such as contaminated soils. While detection of unique intermediates or metabolites is indicative of specific microbial biochemical processes, the results are not unambiguous. The putative metabolic intermediates may be the result of abiotic rather than biotic processes.[11] In labeled mineralization studies, a ^{14}C-substrate is added to a soil slurry in an enclosed flask and the degradative fate of the compound is determined by trapping and analyzing for [14]CO$_2$- and ^{14}C-labeled metabolites. While highly specific, there is some uncertainty regarding the difference in bioavailability between the added labeled compound and the endogenous unlabeled compounds.[7] In addition, activity determinations based on labeled mineralization studies do not correlate well with disappearance studies.[12] If the unlabeled compound is strongly sorbed on soil particles and, therefore, is effectively insoluble, degradative rates based on the bioavailability of weakly sorbed, soluble compounds will be unrealistically high.

NUCLEIC ACID-BASED APPROACHES FOR ASSESSING MICROBIAL ACTIVITY IN SOILS

DNA and Ribosomal rRNA Analysis

Recently the application of nucleic acid hybridization techniques has allowed new approaches to the study of microbial populations and activities *in situ*. Hybridization of gene probes with DNA obtained from environmental samples has been broadly used for microbial identification and the study of microbial population structure and function in the environment.[2-4] These probes are generally cloned fragments of particular genes obtained from organisms cultured in the laboratory. Because of the physical principle on which nucleic acid hybridization is based, that is complimentary nucleotide base pairing between the probe sequence and the target sequence, the use of this technique is highly specific allowing detection and discrimination among extremely heterogeneous sequences. As applied to the remediation of contaminated soils, the gene frequency of the catabolically relevant degradative pathways may be determined by the genetic probing of DNA obtained from soil microorganisms.[13-15] Information about the catabolic genotype may be obtained by probing colony lifts made from plating a soil dilution series on nonselective agar media or by probing the DNA isolated from the soil microorganisms. Probing of isolated DNA avoids the biasing introduced by culturing of environmental microbes, most of which are not culturable, thus giving a better representation of the natural population.[13] DNA may be obtained from cells after first separating them from soil particles by differential centrifugation[16] or, alternatively, by lysing the cells directly in the soil matrix.[13] Both procedures have their advantages and disadvantages.[16] The cell separation procedure is unable to dislodge cells strongly bound to soils particles, and thus results in a lower yield of DNA. However, the DNA so obtained is uncontaminated by soil matrix components. On the other hand, the direct extraction procedure results in higher DNA yields more representative of the actual soil population, but the DNA is contaminated with soil humic acids. These humic materials are amorphous high-molecular-weight polymers of phenolic compounds. Their physical properties, such as charge and molecular weight, are similar enough to nucleic acids so that procedures conventionally used for nucleic acid purification result in the copurification of humic materials.[17] Humic acid contamination has been reported to interfere with enzymatic manipulation of nucleic acids such as restriction analysis.[18] While DNA probing studies are quite valuable in the determination of the population genotype, DNA analysis provides no information as to the viability, catabolic activity, or gene expression of those organisms.

A molecular approach that compliments DNA analysis is the use of ribosomal RNA (rRNA) as a hybridization target. Because there may be as many as 10,000 ribosomes/cell, in actively growing cells, rRNA probing permits greater sensitivity over DNA probing in that it permits direct detection to the level of a single cell.[19] Since the rRNA content of a cell is proportional to growth rate, the amount of probe hybridized per unit of biomass could provide an estimate of the metabolic activity of a population.[19] The number of rRNA targets/cell in this method is well suited for *in situ* whole cell probing and recent advances in the use of fluorescent probes allow the concurrent use of multiple probes.[23] Ribosomal RNA probes have generally been used to infer quantitative evolutionary relationships among diverse organisms.[17,18] 16 S rRNAs have regions that are highly conserved between genera and species adjacent to species-specific regions. Primers may be designed to anneal to these conserved regions thus allowing direct RNA sequencing into the species-specific regions.[22] The use of rRNA probes therefore allows the determination of cellular identification and an estimate of cellular activity simultaneously.

Messenger RNA as an Index of Microbial Activity in Soils

Messenger RNA obtained from environmental samples may also be used as a target to study microbial activity under environmental conditions.[1,2] Because all studied bacterial catabolic genes of environmental importance are regulated at the transcriptional level, this procedure allows the study of gene expression *in situ*. Our laboratory is particularly interested in the problems associated with the enhancement of microbial catabolic activity related to the bioremediation of soils and we have therefore attempted to use the quantitation of catabolic mRNA levels as a means of assessing microbial degradative rates.

The degradative genetic system chosen for this set of experiments was the NAH7 plasmid from the soil bacteria *Pseudomonas putida*, which is capable of completely degrading naphthalene and surviving on it as a sole carbon source.[24] This pathway is composed of two operons, the first of which (*nah*) degrades naphthalene to salicylate, and the second of which (*sal*) converts salicylate to acetaldehyde and pyruvate.[25] The pathway is expressed at a low level without induction; in the presence of naphthalene, salicylate is formed, which induces both operons by interaction with the *nahR* regulatory protein bound to up-stream promoter sequences.

The feasibility of obtaining mRNA from soil microorganisms was first tested using inoculated soils. Pure cultures of *P. putida* G7 cells were used to inoculate both uncontaminated and polyaromatic hydrocarbon-contaminated soils and, subsequently, the soils were processed to obtain mRNA.[26] Soil bacteria were lysed *in situ* using a variation of the hot phenol technique. The total soil bacterial RNA so obtained was then analyzed by northern blotting using a nick-translated *nahABCD* gene probe. The half-life of the *nahA* gene was determined to be 11 min, making it a reasonable target for study.[27] Considering the general difficulty with ribonuclease contamination associated even with isolation of mRNA from pure culture,[28] we were pleased to find that mRNA could be obtained in undegraded form from a soil matrix. At that time we were also able to demonstrate the probing of *nah* mRNA from uninoculated contaminated soils with population densities of 10^8 cells/g soil.[26] Clearly, the sensitivity of this procedure would have to be improved for application to soils with environmentally realistic cell densities on the order of 10^5–10^6 cells/g soil. Since our first report other groups have successfully extracted mRNA from inoculated soils.[29-31] While a number of different mRNA isolation procedures have been applied to soils, including those developed for eukaryotic systems,[29] all appear to give similar extraction efficiencies. Attempts to optimize mRNA extraction by varying buffer molarity or buffering pH did little to enhance mRNA recoveries from soils.[1] We prefer the hot phenol extraction method, because it simultaneously denatures ribonucleases and allows partitioning of soil organic contaminants away from the aqueous phase.

Detection and Quantitation of Soil-Extracted Messenger RNA

To extend the usefulness of the procedure to soil samples with lower cell densities, a more sensitive method of RNA analysis was required. In addition to sensitivity, we wanted a method that would allow absolute quantitation of the isolated mRNA. Slot blots are a possible alternative to northern blots in that the apparatus permits the concentration of the target RNA into a region of approximately 10 mm^2 and is, therefore, more sensitive; slot blots approach a sensitivity of 0.1 pg of RNA. While both northern and slot blots are amenable to quantitation using RNA standards, northern blots have an advantage in that the electrophoresis step allows separation of the total RNA on the basis of molecular weight. This separation process eliminates possible false-positive artifacts, such as the hybridization with rRNA instead of mRNA when using RNA probes, that may be encountered with slot blots.

A recently developed procedure for mRNA analysis is the adaptation of the polymerase chain reaction for the amplification of mRNA. This procedure requires that complimentary DNA (cDNA) first be made from the target mRNA using the enzyme reverse transcriptase followed by amplification of the cDNA by conventional PCR.[32] While this method is exquisitely sensitive, allowing amplification of mRNA from a single cell in pure culture, the very exponential amplification that makes the procedure so powerful makes quantitation of the original mRNA target difficult. The most rigorous PCR quantitation procedure, termed competitive PCR, involves coamplification of an internal control DNA sequence. In the case of mRNA–PCR a control RNA must be generated *in vitro* from a transcription vector containing the target sequence.[33,34] Since the starting concentration of the internal control is known, the amount of the unknown target sequence may be inferred. However, the added difficulty of amplifying internal controls for each sample makes the procedure ungainly for routine screening of multiple samples; the competitive PCR method requires that approximately ten control samples be amplified for each unknown allowing titration of the unknown against a dilution series of competitor. This technique holds great promise for the analysis of low abundance mRNA and is therefore potentially useful for low cell density environmental samples.[35] However, several technical difficulties must be overcome before PCR may be used to quantitate mRNA extracted from soil microorganisms. As with DNA extracted directly from soils, RNA extracted from soil microorganisms is always similarly contaminated. Though amplification of mRNA samples obtained from soil microorganisms has been demonstrated,[31] it seems likely that humic contaminants interfere with both the reverse transcriptase and DNA polymerase steps of the PCR procedure. Adding an amplifiable RNA control to a soil-extracted RNA that may not be amplifiable defeats the purpose of the internal control. In addition, the large number of amplifications/sample required for competitive PCR makes the procedure cumbersome for routine screening.

Another RNA detection method that is both sensitive and rigorously quantifiable is the ribonuclease protection assay (RPA). This procedure involves hybridization of the total extracted RNA of a sample with an antisense *in vitro*-synthesized RNA probe that is complimentary to the target RNA molecule.[36] The resulting RNA–RNA hybrid is subjected to ribonuclease digestion, which specifically cleaves any single-stranded RNA—that is, any RNA that is not identically complimentary and thus hybridized to the antisense probe. After ribonuclease treatment the protected double-stranded RNA fragment is electrophoresed on a gel and may be quantified with protected fragments created by hybridizing known amounts of *in vitro*-synthesized sense-strand RNA with the antisense probe. Quantitation is accomplished relatively easily if the RNA transcription vector described above has two different RNA polymerase promoters on either side of the inserted sequence; an antisense transcript used for the labeled probe may be generated from one promoter, and the sense transcript generated from the other promoter is used to make the standard curve.

The RPA has several advantages over other RNA detection methods. The solution hybridization step is more tolerant of humic acid contamination compared with enzymatic manipulation; humic acids bound to soil-extracted mRNA do not interfere with the RPA. Depending on the specific activity of the antisense probe,[36] the RPA procedure is sensitive to 0.1 pg of mRNA target and, because of the separation of protected fragments on the basis of molecular weight, avoids possible hybridization artifacts. Because ribonuclease digestion of a double-stranded RNA hybrid is sensitive to a single base difference, this procedure should also allow the mapping of community complexity and pathway heterogeneity; minor differences in sequence between similar genes will give multiple protected fragments on a separating gel. If the target transcripts are abundant enough to be detected, this procedure is, therefore, well suited for the analysis of environmental RNA samples in that it permits direct absolute quantitation, is tolerant of environmentally contaminated RNA samples, and is suitable for the screening of multiple samples.

Quantitation of Messenger RNA Extracted Directly from Polyaromatic Hydrocarbon-Contaminated Soils

Four polyaromatic hydrocarbon contaminated soils from a Manufactured Gas Plant site[7] and a creosote site were chosen for naphthalene mRNA analysis. These soils are heavily contaminated with naphthalene and an array of other organic and inorganic compounds.[34] The total heterogeneous bacterial populations were determined with and without an 18-hr incubation in sterile water with shaking at 27°C (Table 4.1). The incubated soils showed a significant increase in growth over the nonincubated soils ($p = 0.05$). Growth was determined on nonselective YEPG spread plates; incubated populations ranged from $1.2 \pm 0.2 \times 10^7$ to $3.4 \pm 0.6 \times 10^8$ cfu/g soil (Table 4.1). Colonies were transferred from agar petri plates to nylon membranes and probed for the naphthalene dioxygenase gene $nahA$, the first enzyme in the naphthalene pathway. The incubated $nahA$-positive populations ranged from $4.0 \pm 1.73 \times 10^4$ to $1.2 \pm 0.3 \times 10^8$ cfu/g soil representing 1–35% of the total bacterial populations. A positive correlation was found between soil naphthalene and $nahA$ gene frequency (for a linear curve, $r^2 = 0.846$, linear; for an exponential curve, $r^2 = 0.818$) (Fig. 4.1). Naphthalene mineralization rates were determined for the four soils by following the catabolites of added ^{14}C-labeled naphthalene. Soil slurries were incubated with shaking for 72 hr with labeled naphthalene in sets of 25-mL vials with NaOH CO_2 traps. At 2 hr intervals for the first 12 hr and at 4 hr intervals thereafter, H_2SO_4 was added to the slurries to kill the cells and the amount of $^{14}CO_2$ produced was determined by scintillation counting. Each of the four soils was catabolically active as determined by [^{14}C]naphthalene mineralization (Table 4.1) and the mineralization rates, as determined on the basis of the calculated water-soluble component of the total soil naphthalene,[38] ranged from 3.2×10^{-5} to 7.8×10^{-1} μg naphthalene/g soil/hr (Table 4.1).

For total RNA extraction 10 g of each soil was suspended in 5 mL of deionized water, incubated overnight with shaking at 27°C, after which total bacterial RNA was extracted directly from the soil slurries using a variation of the hot phenol technique. RNA was also isolated from duplicate soil slurries that had been further incubated with 5 mM salicylate for 30 min. The slurries were

Table 4.1 Comparison of Total Bacterial Populations, *nah*-Positive Bacterial Populations, Percent [^{14}C]1-Naphthalene Mineralized, [^{14}C]1-Naphthalene Rates, and *In Situ* RNA Quantitation from MGP and Creosote-Contaminated Soils[a]

Soil	Unincubated bacterial population[b]	Incubated[a] bacterial population[b]	*nahA* colony hybridization (cfu/g)	^{14}C recovery		[^{14}C]Naphthalene mineralization rate (μg/g soil/hr)	Soluble naphthalene (μg/mL)	^{32}P-labeled *phlA* probe recovery (%)	*nahA* mRNA (pg/g)
				%CO$_2$	% Total				
B	$2.9 \pm 0.6 \times 10^6$	$1.2 \pm 0.2 \times 10^7$	$4.0 \pm 1.7 \times 10^4$	33.8	64	1.8×10^{-4}	8.0×10^{-4}	19.4	—
B/i								2.0	—
C	$1.1 \pm 0.3 \times 10^4$	$2.3 \pm 0.2 \times 10^7$	$3.7 \pm 0.9 \times 10^6$	17.5	64	5.1×10^{-2}	7.9×10^{-2}	3.0, 6.0	24,[d] 37[c]
C/i								1.6	20[c]
D	$1.1 \pm 0.5 \times 10^7$	$1.8 \pm 0.2 \times 10^8$	$1.2 \pm 0.6 \times 10^6$	73.0	92	3.2×10^{-5}	2.0×10^{-3}	2.4	2[c]
D/i								10.5	3[c]
G	$1.6 \pm 0.7 \times 10^7$	$3.4 \pm 0.6 \times 10^8$	$1.2 \pm 0.3 \times 10^8$	81.1	—	7.8×10^{-1}	2.9×10^{-1}	50.0, 12	62[c], 75[d]
G/i								6.2	52[d]

[a] Soils subjected to salicylate induction are indicated by "i" following the soil letter designation. Because final mRNA quantitation depends both on the percent recovery and the RPA value, results are tabulated as individual observations.

[b] Soils incubated in water for 18 hr prior to enumeration.

[c] Final soil bacterial RNA purification by CsCl centrifugation.

[d] Final soil bacterial RNA purification by anion-exchange chromatography.

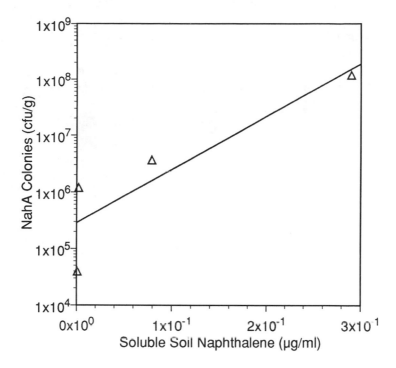

Figure 4.1 Correlation between *nahA*-positive colonies and soluble soil naphthalene (linear, r^2 = 0.846; exponential, r^2 = 0.818).

mixed with 20 mL of 50 mM sodium acetate, 50 mM sodium chloride, 5 mM EDTA, 2% sodium dodecyl sulfate, pH 5.2 and 20 mL phenol/chloroform at 60°C for 10 min and shaken for 5 min with a wrist action shaker. To allow estimation of the percent *nahA* mRNA recovery from the soil slurries, a synthetic labeled mRNA molecule was created *in vitro* and added to the soils at the beginning of the isolation procedure. The synthetic RNA molecule was produced by transcription from a plasmid that contained a 1.4-kb sequence from a nonhomologous phenol hydroxylase *phlA* gene[35] (chosen because it showed no hybridization with *nahA*). The slurries were centrifuged, reextracted with phenol/chloroform, and the nucleic acids precipitated with EtOH. The pellets were resuspended in buffer and applied to ion-exchange columns (Qiagen p-100) and the RNA fractions eluted. The RNA was DNase treated and the percentage of recovered ^{32}P-labeled *phlA* RNA determined by scintillation counting.

A 558-bp fragment of the *nahA* gene was inserted into a transcription vector (pJF12) allowing *in vitro* synthesis of both sense and antisense *nahA* transcripts from opposable T3 and T7 promoters. Purified soil bacterial RNA preparations were allowed to hybridize with a ^{32}P-labeled antisense *nahA* transcript. A standard curve was created by hybridizing *in vitro*-transcribed *nahA* sense strand RNA with antisense ^{32}P-labeled transcripts (Fig. 4.2). After RNase digestion the samples were separated on urea-acylamide gels (Fig. 4.2) and the

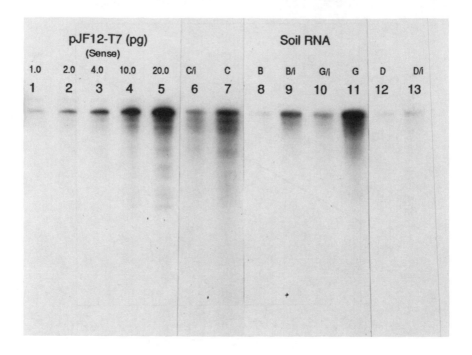

Figure 4.2 Electrophoresis of soil mRNA ribonuclease protection assay (RPA) products. Lanes 1–5, standard curve formed with 1, 2, 4, 10, and 20 pg of *nahA* sense strand transcript. Lanes 6–13, soil bacterial RNA isolated from uninduced and salicylate-induced soils C_i, C (10 g), B, B_i (10 g), G_i (1 g), G (2 g), and D, D_i (10 g) purified by ion-exchange chromatography.

protected bands corresponding to the molecular weight of the sense-strand-protected fragment were quantified by comparison against the standard curve. The RPA results were normalized individually on the basis of the percent recovery of the labeled probe RNA from each extraction to give the *nahA* mRNA/g soil, the RPA (Table 4.1). While labeled RNA recoveries were variable ranging from 1.6 to 50%; calculated *nahA* mRNA levels were comparable in replicate experiments (soils C and G, Table 4.1). *NahA* transcript levels ranged from 2 to 75 pg/g soil. The soil slurry conditions of the incubated soils approximated those of the mineralization study, thus permitting comparison of the results from the mRNA extraction, population enumeration, and mineralization experiments. The concentration of soil naphthalene correlated well with uninduced soil *nahA* mRNA levels ($r^2 = 0.919$) (Fig. 4.3) and [^{14}C]naphthalene mineralization rates ($r^2 = 0.957$) (Fig. 4.4).

The four soils showed no induction by salicylate. While preliminary experiments suggested that soil B was inducible, these findings were not reproducible. The *nah*-positive population of soils C, D, and G, which had 7.9×10^{-2}, 2.0×10^{-3}, and 2.9×10^{-1} μg soluble naphthalene/mL, respectively, may have been fully induced. An alternative explanation for the lack of inducibility may be the presence of inhibitory contaminants; besides PAHs these soils contain

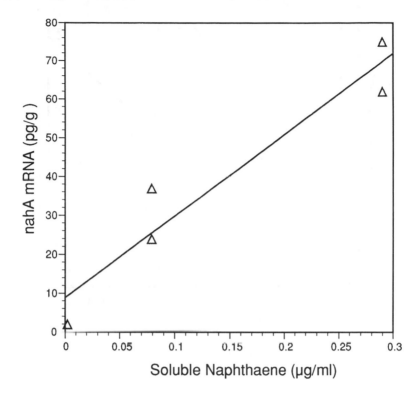

Figure 4.3 Correlation between soluble soil naphthalene concentrations and uninduced soil bacterial *nahA* mRNA levels (r^2 = 0.919).

significant concentrations of inorganic compounds and metals. It is reasonable that while transcription of the *nahA* gene was not inhibited, perhaps transcription of the other genes in the pathway were inhibited or that the pathway enzymes were in fact translated but the enzymes were inhibited. However, the fact that *nahA*-positive colonies increase in proportion to the naphthalene concentration (Fig. 4.1) argues against this hypothesis; bacteria that are predominantly *nah* positive are able to thrive in PAH-contaminated soil. In addition, all the soils showed bacterial growth after incubation in sterile water (Table 4.1) suggesting that soil nutrients such as nitrogen or phosphorous were not limiting.

CONCLUSIONS

While the correlation of a single transcript with the integrated activities of an entire catabolic pathway seems unlikely, the observed correlation between *nahA* mRNA and ^{14}C mineralization rates suggests that transcriptional analysis may be a valid index for catabolic activity in PAH-contaminated soils. Recovery of *nahA* mRNA from soil B, while initially successful, was not reproducible,

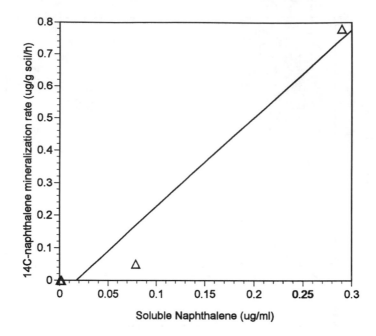

Figure 4.4   Correlation between soluble soil naphthalene concentrations and [^{14}C]1-naphthalene mineralization rates (r^2 = 0.957).

suggesting that a soil population of greater than 4×10^4 inducible cells/g soil is required for rigorous quantitation of mRNA by the RPA.

ACKNOWLEDGMENTS

This work was supported by Electric Power Research Institute Contract RP-3015-1 and partial support by Gas Research Institute Contract 5089-226-1933 and United States Air Force Office of Scientific Research Contract F-49620-89-C-0023. The technical assistance of J. Sanseverino and B. Applegate is gratefully acknowledged.

REFERENCES

1. Sayler, G. S., Fleming, J. T., Applegate, B., and Werner, C. Nucleic acid extraction and analysis: Detecting genes and their activity in the environment. *Genetic Interactions Between Microorganisms in the Natural Environment*. Wellington, E. M. and VanElas, J. D., Eds. Manchester Press, Manchester, UK, 1991, p. 233.
2. Sayler, G. S., Nikbakht, K., Fleming, J. T., and Packard, J. Application of molecular techniques to soil, biochemistry. *Soil Biochem.* 7, 131, 1991.

3. Atlas, R. M., Sayler, G. S., Burlage, R. S., and Bej, A. K. Molecular approaches for environmental monitoring of microorganisms. *Biotechniques* 12, 706, 1992.
4. Jain, R. K., Burlage, R. S., and Sayler, G. S. Methods for detecting recombinant DNA in the environment. *CRC Crit. Rev. Biotechnol.* 8, 33, 1988.
5. Blackburn, J. W., Troxler, W. L., and Sayler, G. S. Prediction of the fates of organic chemicals in a biological treatment process—an overview. *Environ. Prog.* 3, 163, 1987.
6. Sayler, G. S. and Fox, R. Environmental biotechnology for waste management. *Environmental Science Research.* Sayler, G. S., Fox, R., and Blackburn, J. W., Eds. Plenum Press, New York, 1991, Vol. 41, p. 1.
7. Blackburn, J. W., DiGrazia, P. M., and Sanseverino, J. Treatability and scale-up protocols for polynuclear aromatic hydrocarbon bioremediation of manufactured gas plant soils. Gas Research Institute Report No. 5087-253-1490. Gas Research Institute, Chicago, 1991.
8. Greenberg, A. E., Trussell, R. R., Clesceri, L. S., and Franson, M. H., Eds. *Standard Methods for the Examination of Water and Wastewater,* 16th ed., American Public Health Association, Washington D.C., 1985, p. 525.
9. Webster, J. J., Hall, M. S., and Leach, F. R. ATP and adenylate energy charge determinations on core samples from an Av-fuel spill site at the Travers City, Michigan airport. *Bull. Environ. Contam. Toxicol.* 49, 232, 1992.
10. Stevenson, L. H., Chrzanowski, T. H., and Erkenbrecher, C. W. The adenosine triphosphate assay: Conceptions and misconceptions. *Native Aquatic Bacteria: Enumeration, Activity and Ecology.* Costerton, J. W. and Colwell, R. R., Eds. American Society for Testing and Materials, Philadelphia, 1979, p. 99.
11. Madsen, E. L. Determining in situ biodegradation. *Environ. Sci. Technol.* 25, 1663, 1991.
12. Rochkind-Dubinsky, M. L., Sayler, G. S., and Blackburn, J. W. *Microbial Decomposition of Chlorinated Aromatic Compounds.* Marcel Dekker, New York, 1987, p. 39.
13. Ogram, A., Sayler, G. S., and Barkay, T. The extraction and purification of microbial DNA from sediments. *J. Microbiol. Methods.* 7, 57, 1987.
14. Ogram, A. V. and Sayler, G. S. The use of gene probes in the rapid analysis of natural microbial communities. *J. Indust. Microbiol.* 3, 281, 1988.
15. Sayler, G. S., Shields, M. S., Breen, A., Tedford, E. T., Hooper, S., Sirotkin, K. M., and Davis, J. W. Application of DNA-DNA colony hybridization to the detection of catabolic geneotypes in environmental samples. *Appl. Environ. Microbiol.* 48, 1295, 1985.
16. Steffan, R. J., Goksoyr, J., Bej, A. K., and Atlas, R. M. Recovery of DNA from soils and sediments. *Appl. Environ. Microbiol.* 54, 2908, 1988.
17. Andreux, F. Genesis and properties of humic moleucles. *Constituents and Properties of Soils.* Bonneau, M. and Souchier, B., Eds. Academic Press, New York, 1982, p. 109.
18. Holben, W. E., Jansson, J., Chelm, B., and Tiedje, T. DNA probe method for detection of specific microorganisms in the soil bacterial community. *Appl. Environ. Microbiol.* 54, 703, 1988.
19. Woese, C. R., Stackebrandt, E., Macke, T. J., and Fox, G. E. A phylogenetic definition of the major eubacterial taxa. *Sys. Appl. Microbiol.* 6, 143, 1985.
20. Olson, G. J., Lane, D. J., Giovannoni, S. J., and Pace, N. R. Microbial ecology and evolution: A ribosomal RNA approach. *Annu. Rev. Microbiol.* 40, 337, 1986.

21. Stahl, D. A., Sogin, M. L., and Pace, B. Rapid determination of 16S ribosomal RNA sequences for phylogenetic analysis. *Proc. Natl. Acad. Sci. U.S.A.* 82, 6955, 1985.

22. Giovannoni, S. J., DeLong, E. F., Olsen, G. J., and Pace, N. R. Phylogenetic group-specific oligodeoxynucleotide probes for identification of single microbial cells. *J. Bacteriol.* 170, 720, 1988.

23. Ried, T., Baldini, A., Rand, T. C., and Ward, D. C. Simultaenous visualization of seven different DNA probes by in situ hybridization using combinatorial fluorescence and digital imaging microscopy. *Proc. Natl. Acad. Sci. U.S.A.* 89, 1388, 1992.

24. Yen, K. M. and Serdar, C. M. Genetics of naphthalene catabolism in Pseudomonads. *CRC Crit. Rev. Microbiol.* 15, 247, 1988.

25. Dunn, N. and Gunsalus, I. C. Transmissible plasmids coding early enzymes of naphthalene oxidation in *Pseudomonas putida*. *J. Bacteriol.* 114, 974, 1973.

26. Sayler, G. S., Fleming, J. T., Applegate, B., Werner, C., and Nikbakht, K. Microbial community analysis using environmental nucleic acid extracts. *Recent Advances in Microbial Ecology.* Hattori, T., Ishida, Y., Maruyama, Y., Morita, R., and Uchida, A., Eds. Japan Scientific Societies Press, Tokyo, 1989, p. 658.

27. Unpublished observations.

28. Blumberg, D. D. Creating a ribonuclease-free environment. *Guide to Molecular Cloning Techniques.* Berger, S. L. and Kimmel, A. R., Eds. Academic Press, San Diego, CA, 1987, p. 20.

29. Tsai, Y., Park, M. J., and Olson, B. H. Rapid method for direct extraction of mRNA from seeded soils. *Appl. Environ. Microbiol.* 57, 765, 1991.

30. Ogunseitan, O. A., Delgado, I. L., Tsai, Y. L., and Olson, B. H. Effect of 2-Hydroxybenzoate on the maintenance of naphthalene degrading pseudomonads in seeded and unseeded soil. *Appl. Environ. Microbiol.* 57, 2873, 1991.

31. Selenska, S. and Klingmuller, W. Direct recovery and molecular analysis of DNA and RNA from soil. *Microb. Releases* 1, 41–46, 1992.

32. McCabe, P. C. Production of single stranded DNA by asymmetric PCR. *PCR Protocols: A Guide to Methods and Applications.* Innis, M.A., Gelf, D. H., Sninsky, J. J., and White, T. J., Eds. Academic Press, San Diego, CA, 1990, p. 76.

33. Wang, A.M., Doyle, M.V., and Mark, D. Quantitation of mRNA by the polymerase chain reaction. *Proc. Natl. Acad. Sci. U.S.A.* 86, 9717, 1989.

34. Gilliland, G., Perrin, S., Blanchard, K., and Bunn, K. F. Analysis of cytokine mRNA and DNA: Detection and quantitation by competitive polymerase chain reaction. *Proc. Natl. Acad. Sci. U.S.A.* 87, 2725, 1990.

35. Steffan, R. J. and Atlas, R. M. Polymerase chain reaction: Applications in environmental microbiology. *Annu. Rev. Microbiol.* 45, 137, 1991.

36. Melton, D. A., Krieg, P. A., Rebagliati, M. R., Maniatis, T., Zinn, K., and Green, M. R. Efficient *in vitro* synthesis of biologically active RNA and RNA hybridization probes from plasmids containing a bacteriophage SP6 promoter. *Nucleic Acid Res.* 12, 7035, 1984.

37. Cushy, M. A. and Morgan, D. J. Biological treatment of soils contaminating manufactured gas plant residues. Gas Research Institue Topical Report No. GRI-90/0117. Gas Research Institute, Chicago, 1990.

38. Sanseverino, J., Werner, C., Fleming, J. T., Applegate, B., King, J. M. H., and Sayler, G. S. Molecular diagnostics of polycyclic aromatic hydrocarbon biodegradation in manufactured gas plant soils. *Biodegradation* 4, 303–321, 1992.

39. Kukor, J. J. and Olsen, R. H. Molecular cloning, characterization and regulation of a *Pseudomonas picketti* PK01 gene encoding phenol hydroxylase and expression of the gene in *Pseuodomonas aerginosa* PA01c. *J. Bacteriol.* 172, 4624, 1990.

CHAPTER 5

Biological and Biochemical Evidence of Strand-Specific Repair of DNA Damage Induced in Human Cells by (±)-7β,8α-Dihydroxy-9α,10α-Epoxy-7,8,9,10-Tetrahydrobenzo[a]pyrene

Veronica M. Maher, Ruey-Hwa Chen, and J. Justin McCormick

INTRODUCTION

To investigate the mechanisms by which normal human cells are transformed into malignant cells, we have been investigating the relationship of DNA repair and DNA replication to mutagenesis induced in diploid human fibroblasts by carcinogens. We showed previously that the frequency of 6-thioguanine (TG)-resistant mutants induced in normal human cells by (±)-7β,8α-dihydroxy-9α,10α-epoxy-7,8,9,10-tetrahydrobenzo[a]pyrene (BPDE),[1] is significantly lower than it is in xeroderma pigmentosum (XP) cells that are virtually devoid of nucleotide excision repair capacity, i.e., XP12BE cells from complementation group A. We synchronized populations of normal or XP cells by release from confluence, and showed that the frequency of mutants induced by BPDE is very high in normal cells treated at the beginning of S phase, when the target gene for mutations, i.e., hypoxanthine(guanine)phosphoribosyltransferase (*HPRT*) is being replicated, and much lower if they are treated in early G_1 phase 12 hr or more prior to the onset of S phase.[1] No such decrease was seen when XP12BE cells were used. These results suggest that semiconservative DNA replication during S phase converts the covalently bound BPDE residues (DNA adducts) into mutations, and that normal cells have time to remove such adducts before S phase replication. If so, they do so in an error-free manner.

BPDE forms adducts principally with guanine bases at the N^2 position.[2] We wanted to determine the nature of the mutations induced in the coding region of the *HPRT* gene in normal cells and in the XP12BE cells exposed in early S

0-87371-951-4/95/$0.00+$.50

or G_1 phase of the cell cycle, to see if the mutations are targeted to guanine · cytosine base pairs, and also to determine the effect of excision repair on the kinds of mutations and their location in the gene (spectrum).

Bohr et al.[3] showed that in a hamster cell line, UV-induced pyrimidine dimers are excised more rapidly from an actively transcribed gene, dihydrofolate reductase (dhfr), than from the genome overall, or from the 5' or 3' flanking region of the gene.[4] Mellon et al.[5] showed that such preferential repair of active genes also occurs in human cells, and that these UV-induced lesions are removed from the transcribed strand of the DHFR gene much more rapidly than from the nontranscribed strand (strand-specific repair).[6]

Human cells remove BPDE adducts using nucleotide excision repair. If one assumes that guanine adducts are the premutagenic lesions induced by BPDE, and that these are formed randomly in the gene, such adducts should be distributed in a ratio 38% in the transcribed strand: 62% in the nontranscribed strand. (This is the ratio of guanine nucleotides able to affect an amino acid in the coding region of the HPRT gene.) If nucleotide excision repair-proficient diploid human fibroblasts are synchronized and irradiated just as the HPRT gene is being replicated, the premutagenic guanine lesions responsible for the mutations observed should be located in both strands. Conversely, if the cells are irradiated in G_1 phase and allowed several hours for excision repair before the HPRT gene is replicated, the premutagenic photoproducts ought to be located predominantly in the nontranscribed strand. No such switch in strand distribution of premutagenic lesions should be seen in the excision repair-deficient XP12BE cells.

STUDIES OF THE SPECTRA OF MUTATIONS INDUCED BY BPDE

To test this hypothesis we wanted to determine the nature of the mutations induced by BPDE in the HPRT gene of these human cells and their location in the gene. Because finite life span human cells that have been cloned, e.g., selected for TG-resistance, cannot easily be expanded to large populations of progeny cells, Yang et al.[7] worked out conditions to allow us to transcribe mRNA into cDNA directly from the lysate of a small clone of cells and then amplify the cDNA of the HPRT gene 10^{11}-fold and sequence the double-stranded DNA product directly. We used this technique to investigate the cell cycle dependence of the strand distribution of premutagenic BPDE adducts in the two populations treated at S or in G_1 phase.

Studies with Repair-Proficient Cells

Chen et al.[8] synchronized populations of normal human fibroblasts, exposed them to BPDE (0.16–0.23 μM) in early S phase or in early G_1 phase, and assayed them for survival and for the frequency of cells resistant to 6-thioguanine (TG). (TG selects for cells lacking HPRT activity.) There was a dose-dependent linear

Table 5.1 Types of Base Substitutions Induced by BPDE in the Coding Region of the *HPRT* Gene in Repair-Proficient Cells and Repair-Deficient XP12BE Cells Treated in S Phase and in G_1 Phase

	Number of substitutions observed			
Type of base substitution	Repair-proficient cells		XP12BE fibroblasts	
	S phase	G_1 phase	S phase	G_1 phase
Transversions:				
$G \cdot C \rightarrow T \cdot A$	16	13	12	15
$G \cdot C \rightarrow C \cdot G$	2	3	2	3
$A \cdot T \rightarrow C \cdot G$	0	0	1	0
$A \cdot T \rightarrow T \cdot A$	0	0	0	2
Transitions:				
$G \cdot C \rightarrow A \cdot T$	1	3	4	3
$A \cdot T \rightarrow G \cdot C$	1	0	1	1
Total	20	19	20	24

Data for this table are taken from references 8 and 9.

increase in mutant frequency in both populations, but the frequency induced in populations treated in early G_1 phase was 3-fold lower than that of cells treated in early S. At a dose of 0.23 μM, which lowered the survival of the cells to 25% of the untreated control cells, the frequency in the S phase population was 300 $\times 10^{-6}$; in the G_1 phase population it was 90 $\times 10^{-6}$. To decrease the chances of including a mutant containing a spontaneous mutation, rather than a BPDE-induced mutation, we sequenced mutants derived only from experiments in which the induced mutant frequency was at least 10 times background.

Examination of the sequencing data from 23 TG resistant clones from cells treated in S phase and 23 from cells treated in G_1 phase showed that the majority were base substitutions. There was no significant difference between S and G_1 phase-derived mutants in the kinds of mutations they contained.[8] There also was no significant difference in the kinds of base substitutions observed in the mutants obtained from the two populations (Table 5.1). All except one S phase-derived mutant in the normal cells involved $G \cdot C$ base pairs. Transversion of $G \cdot C$ base pairs to $T \cdot A$ base pairs was the most frequent base substitution mutation, 80% for S phase and 68% for G_1 phase.[8]

Strand Distribution of the Premutagenic Lesions

Because >95% of the DNA adducts induced by BPDE involve guanine, the mutation spectrum data allow one to determine which strand contained the premutagenic lesions that gave rise to the observed $G \cdot C$ base substitution mutations. In mutants from cells treated in S phase, the premutagenic guanines were located in both strands, i.e., 24% were in the transcribed strand and 76% in the nontranscribed strand. The distribution predicted from the ratio of guanines that, if changed, would affect the amino acid composition of the protein is 38% transcribed:62% nontranscribed. (The distribution of premutagenic guanine lesions in the XP12BE cells was also 24% transcribed

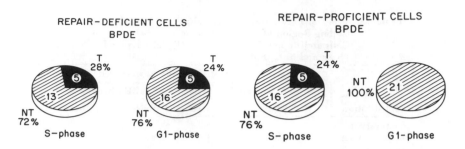

Figure 5.1 Diagram of the strand distribution of the premutagenic BPDE lesions.

strand, 76% nontranscribed strand, see Fig. 5.1.) In mutants from normal cells treated in G_1 phase and allowed at least 12 hr for excision repair before DNA synthesis began, 100% of the premutagenic guanines were located in the nontranscribed strand. This difference is statistically significant ($p < 0.05$, using the chi square test) and supports the hypothesis that BPDE adducts are excised in a strand-specific manner.

Studies with Repair-Deficient Cells

Similar studies were carried out with synchronized populations of XP12BE cells treated in S or early G_1 phase but using much lower doses of BPDE (0.013–0.022 μM). In both sets of XP12BE cells, the frequency of TG-resistant cells per 10^6 clonable cells ranged from 25 to 94, with a background frequency of 5×10^{-6}. To decrease the possibility of including a background mutant, we sequenced mutants only if they were derived from populations in which the frequency was at least 8 times higher than background. A total of 25 S phase-derived mutants and 28 G_1 phase-derived mutants were analyzed. We found no significant difference between S- and G_1-derived mutants in the kinds of mutations seen. The majority contained base substitutions. Of the base substitutions derived from either phase 89% involved $G \cdot C$ base pairs, with transversions of $G \cdot C$ to $T \cdot A$ predominating; 60% (12/20) for S phase and [62% (15/24) for G_1 phase] (Table 5.1). As shown in Figure 5.1, in mutants from S phase, five of the guanines involved in a base substitution out of 18 (28%) were located in the transcribed strand. Similarly, in those mutants derived from G_1 phase, five of the guanines out of 21 (24%) were located in the transcribed strand.[9] These results are equivalent to those seen with normal cells treated at S phase.

EXPLANATION FOR THE STRAND BIAS IN DISTRIBUTION OF PREMUTAGENIC GUANINES

These results support the hypothesis that nucleotide excision repair of BPDE-induced adducts occurs preferentially on the transcribed strand. If one assumes that binding of BPDE to guanine in the *HPRT* gene is random, and virtually every amino acid change can be detected because it renders a cell resistant to thioguanine, in cells incapable of excision (XP12BE) and in repair-proficient cells that had almost no time for excision before the DNA replication fork encountered the BPDE adduct (normal cells treated in early S), 38% of the mutations seen should have corresponded to a guanine in the transcribed strand. Only 24% did. We attribute this bias (from 38 to 24%) to the effect of silent mutations. In other words, we suggest that in those places in the coding region of the *HPRT* gene where an amino acid change does not affect the activity of the HPRT protein, the guanines are located in the transcribed strand and, therefore, mutations involving these guanines will not be included among the mutations observed unless, by chance, a second mutation (nonsilent) in the same cell allows the mutant to be selected by thioguanine.

Support for this hypothesis comes from the results of McGregor et al.[10] who carried out similar studies using ultraviolet (UV) radiation as the carcinogen (a mutagen that primarily induces base substitutions involving $G \cdot C$ base pairs, in which the pyrimidine is the premutagenic base). They found that normal cells irradiated in early S phase exhibited a high frequency of mutations; those irradiated in mid G_1 phase, 6 hr prior to the onset of S, exhibited a frequency only half that of the cells irradiated at S, and cells irradiated 11 hr prior to S phase had a frequency near that of background. In contrast, XP12BE cells irradiated with a 10-fold lower dose in early S phase or 11 hr prior to S phase showed a high frequency of mutations in both populations. McGregor et al.[10] sequenced mutants from S phase and mid G_1 phase for the normal cells and S and early G_1 phase for the XP12BE cells. There was no significant difference in the kinds of base substitutions induced in the four populations, but there was a striking difference in the strand distribution of the premutagenic lesions. In the normal cells irradiated in early S, the strand distribution of photoproducts responsible for the observed mutations was 71% transcribed:29% nontranscribed; in normal cells allowed 6 hr for excision repair before S phase, the ratio was 20% transcribed:80% nontranscribed. In the XP12BE cells irradiated both at S phase and in early G_1 phase, the ratio averaged 74%:26%. The distribution of cytosines, which when involved in a base substitution will affect an amino acid in the coding region of the *HPRT* gene, is 62%:38%. (For all dipyrimidine photoproducts, this ratio is 59%:41%.) These results suggest that in the critical places in the *HPRT* gene that involve a $C \cdot G$ base pair, the cytosine is located in the transcribed strand, i.e., just the opposite to what was found for premutagenic guanine bases.

STUDIES ON THE RATE OF EXCISION OF BPDE ADDUCTS

Biochemical Evidence of Strand-Specific Repair

Although the evidence of strand-specific repair of UV-induced cyclobutane pyrimidine dimers (CPD) is well documented, Tang and Zhang,[11] using a

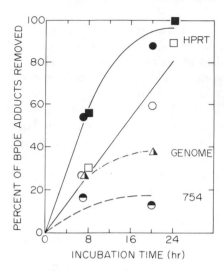

Figure 5.2 Rate of removal of BPDE adducts from the transcribed (closed symbols) and nontranscribed (open symbols) strand of the *HPRT* gene and from both strands of the 754 locus (half closed circles), as determined using UvrABC excinuclease, and from the genome overall (half closed triangles) as determined using loss of tritiated BPDE. Circles indicate data obtained from cells treated with 1.0 µM BPDE; squares and triangles, from cells treated with 1.2 µM BPDE.

variation of the technique developed by Bohr et al.[4] to detect lesions present in specific genes, but using *E. coli* UvrABC excinuclease to recognize and incise bulky chemical adducts, found no difference in the rate of repair of BPDE adducts in the active and nonactive (nontranscribed) regions of the *dhfr* gene of CHO cells, i.e., they were not preferentially removed. This suggested that one might not expect strand-specific repair to occur either. Because our biological data indicated strand-specific repair of BPDE adducts in human cells, we developed that type of assay for the *HPRT* gene to measure the ability of human cells to remove BPDE adducts from either strand of the gene.[12] We also tested for preferential repair of BPDE adducts by comparing the rate of their removal from an active gene, *HPRT*, and an inactive locus of the same chromosome, the 754 locus.

For these studies, we synchronized repair-proficient cells, and treated them with BPDE (1.0 or 1.2 µM) in early G_1 phase so there would be a long period for excision repair before DNA replication. We harvested one set of cells immediately and the rest after various times, and extracted the DNA. Using UvrABC excinuclease, which specifically and quantitatively incises at least 80% of BPDE-DNA adducts of human genomic DNA,[13] and using Southern blotting and hybridization with probes specific for the individual strands of the *HPRT* gene, we quantified the initial number of BPDE adducts formed in the two strands and the number remaining at various times post treatment. The results are shown in Figure 5.2. The rate of removal of BPDE adducts from the transcribed strand was significantly faster than from the nontranscribed strand. Within 7 hr, 53% of the adducts had been removed from the transcribed strand, but only 26% from the nontranscribed strand. By 20 hr, 87% of the adducts had been removed from the transcribed strand, but only 58% from the nontranscribed strand.

Evidence of Preferential Repair of BPDE Adducts in Human Cells

To see if there were a difference in the rate of repair from a nontranscribed (inactive) genomic sequence compared to that from the actively transcribed *HPRT* gene, we measured the formation of such adducts and their rate of repair in the transcriptionally inactive 754 locus. The efficiency of repair in the 754 locus was markedly reduced. During the 20-hr period after treatment, only ~14% of the adducts were removed from this region, compared to 87% and 58% from the transcribed and nontranscribed strands of the *HPRT* gene, respectively (Fig. 2). This rate of repair was even slower than the rate of removal of tritiated BPDE adducts from the overall genome of synchronized cells exposed to 1.2 mM BPDE.

POSSIBLE MECHANISMS FOR STRAND-SPECIFIC REPAIR

The data in Figure 5.2 support our hypothesis that strand-specific repair is responsible for the difference in strand distribution of the premutagenic guanine lesions observed in repair-proficient human cells treated in S phase and such cells treated in G$_1$ phase (Fig. 5.1). Such strand-specific repair suggests a specific coupling between transcription and repair.[5,6] Recently, an assay for such repair by cell-free extracts of *E. coli* was developed, and a candidate "transcription-repair coupling factor" was partially purified from such extracts.[14] This coupling factor could lead to a faster rate of repair in the transcribed strand than in the opposite strand by associating with component(s) of the transcriptional complex via protein–protein interactions and facilitating the assembly of the repair complex in the vicinity of the transcriptional complex, so that repair enzymes can scan the strand that is being transcribed. Another possibility is that this factor directly or indirectly recognizes a unique DNA structure or other signal(s) generated by the stalled transcriptional complex at the site of DNA damage and targets the repair enzymes to the template strand.

Our finding that the removal of BPDE adducts from the transcriptionally inactive 754 locus was much less efficient than from either strand of the *HPRT* gene indicates that there are at least three different rates of repair of removal of BPDE adducts from the genome of human cells. The slow repair rate we observed in the 754 locus agrees with the hypothesis that DNA damage located in condensed, inactive chromatin is less accessible to repair enzymes than that in active chromatin.[5]

ACKNOWLEDGMENTS

The research was supported in whole or in part by DHHS Grant CA21253 from the National Cancer Institute.

REFERENCES

1. Yang, L. L., Maher, V. M., and McCormick, J. J. Relationship between excision repair and the cytotoxic and mutagenic effect of the "anti" 7,8-diol-9,10-epoxide of benzo[a]pyrene in human cells. *Mutat Res.* 94, 435, 1982.
2. Weinstein, I. B., Jeffrey, K. W., Jeanette, K. W., Blobstein, S. H., Harvey, R. G., Harris, C., Autrup, H., Kasai, H., and Nakanishi, K. Benzo[a]pyrene diol epoxides as intermediates in nucleic acid binding *in vitro* and *in vivo*. *Science* 193, 592, 1976.
3. Bohr, V. A., Smith, C. A., Okumoto, D. S., and Hanawalt, P. C. DNA repair in an active gene: Removal of pyrimidine dimers from DHFR gene of CHO cells is much more efficient than in the genome overall. *Cell* 40, 359, 1985.
4. Bohr, V. A., Okumoto, D. S., and Hanawalt, P. C. Survival of UV-irradiated mammalian cells correlates with efficient DNA repair in an essential gene. *Proc. Natl. Acad. Sci. U.S.A.* 83, 3830, 1986.
5. Mellon, I., Bohr, V. A., Smith, C. A., and Hanawalt, P. C. Preferential DNA repair of an active gene in human cells. *Proc. Natl. Acad. Sci. U.S.A.* 83, 8878, 1986.
6. Mellon, I., Spivak, G., and Hanawalt, P. C. Selective removal of transcription-blocking DNA damage from the transcribed strand of the mammalian DHFR gene. *Cell* 51, 241, 1987.
7. Yang, J.-L., Maher, V. M., and McCormick, J. J. Amplification and direct nucleotide sequencing of cDNA from the lysate of low numbers of diploid human cells. *Gene* 83, 347, 1989.
8. Chen, R.-H., Maher, V. M., and McCormick, J. J. Effect of excision repair by diploid human fibroblasts on the kinds and locations of mutations induced by (±)-7β,8α-dihydroxy-9α,10α-epoxy-7,8,9,10-tetrahydrobenzo[a]pyrene in the coding region of *HPRT* gene. *Proc. Natl. Acad. Sci. U.S.A.* 83, 8680, 1990.
9. Chen, R.-H., Maher, V. M., and McCormick, J. J. Lack of a cell cycle-dependent strand bias for mutations induced in the *HPRT* gene by (±)-7β,8α-dihydroxy-9α,10α-epoxy-7,8,9,10-tetrahydrobenzo[a]pyrene in excision repair-deficient human cells. *Cancer Res.* 51, 2587, 1991.
10. McGregor, W. G., Chen, R.-H., Lukash, L., Maher, V. M., and McCormick, J.J. Cell cycle-dependent strand bias for UV-induced mutations in the transcribed strand of excision repair-proficient human fibroblasts, but not in repair-deficient cells. *Mol. Cell. Biol.* 11, 1927, 1991.
11. Tang, M.-S. and Zhang, X.-S. Repair of benzo[a]pyrene diol epoxide-DNA adducts in dihydrofolate reductase gene in Chinese hamster ovary cells. *Proc. Am. Assoc. Cancer Res.* 32, 6, 1991 (abstr).
12. Chen, R.-H., Maher, V. M., Brouwer, J., van de Putte, P., and McCormick, J. J. Preferential repair and strand-specific repair of benzo(a)pyrene diol epoxide adducts from the *HPRT* gene of diploid human fibroblasts. *Proc. Natl. Acad. Sci. U.S.A.* 88, 5413, 1992.

13. Van Houten, B., Masker, W. E., Carrier, W. L., and Regan, J. D. Quantitation of carcinogen-induced DNA damage and repair in human cells with the UVR ABC excision nuclease from *Escherichia coli. Carcinogenesis* 7, 83, 1986.
14. Selby, C. P. and Sancar, A. Gene- and strand-specific repair *in vitro:* Partial purification of a transcription-repair coupling factor, *Proc. Natl. Acad. Sci. U.S.A.* 88, 8232, 1991.

CHAPTER 6

Molecular Analysis of Frameshift Mutations Induced by Plant-Activated 2-Aminofluorene

Michael J. Plewa, Margaret J. Timme, David Cortez, Shannon R. Smith, and Elizabeth D. Wagner

INTRODUCTION

Plant Activation

Plants can activate promutagens into stable mutagens and these genotoxic agents may be hazardous to the environment and to the public health.[1,2] Plant systems have been widely employed in classical and environmental mutagenesis. However, the environmental and human health impact of plants exposed to environmental xenobiotics were not well recognized until the presence of pesticide contaminants in food supplies caused alarm. The capability of plants to bioconcentrate environmental agents and activate promutagens into toxic metabolites is significant when one realizes the immense diversity of xenobiotics to which plants are intentionally and unintentionally exposed. Finally, we all must be attentive to the effects that toxic agents may have on the biosphere and the grave global consequences that would result in a disruption in the carbon cycle.

Plant activation is the process by which a promutagen is metabolically transformed into a mutagen by a plant system. In mammalian systems the majority of enzymes participating in oxidative desulfuration, dealkylation, epoxidation, or ring hydroxylation involve cytochrome P-450-type monooxygenases. Although limited data exist about the inducibility of plant cytochrome P-450, it is unknown if there is an equivalent inducible system to hepatic monooxygenases.[3] Plant peroxidases catalyze the oxidation of a diverse class of xenobiotics. Peroxidases are ubiquitous in plants, however, only limited data are available that demonstrate their participation in the *in vivo* metabolism of foreign compounds.[4-6] The plant activation of 2-aminofluorene has been well studied in our laboratory. It is activated into potent frameshift mutagens by a

0-87371-951-4/95/$0.00+$.50

number of cultured plant cell species as detected by *Salmonella typhimurium* tester strains in the plant cell/microbe coincubation assay.[7,8]

Mutant Spectra

Mutant spectra analysis has been conducted on many of the *S. typhimurium* tester strains.[9-13] These studies define the molecular basis of both spontaneous and induced revertants. We used three strains of *S. typhimurium*, YG1024, TA98, and TA98/1,8-DNP$_6$ to detect *his*$^+$ reversion at the *hisD3052* allele using a plant-activation assay. We conducted mutant spectra analysis with these strains and compared the spontaneous and the plant-activated 2-aminofluorene-induced revertants of YG1024 and TA98.[14,15] In other studies, a high frequency of reversion occurs by a CG/GC deletion that is located in an alternating CG octamer of the *hisD3052* allele (D878–885). A colony-probe hybridization procedure was developed to detect this –2 deletion.[9,13]

The objectives of this research were to construct a mutant spectra of the YG1024 and TA98 revertants induced by the plant-activated 2-aminofluorene metabolite(s) and to compare these spectra with that from spontaneous revertants of the identical strains.

MATERIALS AND METHODS

Plant Cell/Microbe Coincubation Assay

The assay is based on employing living plant cells in suspension culture as the activating system and specific microbial strains as the genetic indicator organism.[7,16] The plant and microbial cells are coincubated together in a suitable medium with a promutagen. The activation of the promutagen is detected by plating the microbial suspension on a selective medium; the viability of the plant and microbial cells may be monitored as well as other components of the assay (Fig. 6.1). Long-term plant cell suspension cultures of tobacco (*Nicotiana tabacum*), cell line TX1, were maintained in MX medium. *S. typhimurium* strains TA98, TA98/1,8-DNP$_6$, and YG1024 were the genetic indicator organisms used. A TX1 cell culture was grown at 28°C to early stationary phase, and the cells were washed and suspended in MX$^-$ medium. MX$^-$ medium lacks plant growth hormone. The fresh weight of the plant cells was adjusted to 100 mg/mL, and the culture was stored on ice (≤30 min) until used. An overnight culture of *S. typhimurium* was grown from a single colony isolate in 100 mL of Luria broth (LB) at 37°C with shaking. The bacterial suspension was centrifuged and washed in 100 mM potassium phosphate buffer, pH 7.4. The titer of the suspension was determined spectrophotometrically at 660 nm and adjusted to 1×10^{10} cells/mL, and the culture was placed on ice. In the coincubation assay, each reaction mixture consisted of 4.5 mL of the plant cell suspension in MX$^-$ medium, 0.5 mL of the bacterial suspension (5×10^9 cells), and a known amount

PLANT/MICROBE COINCUBATION ASSAY

Bacterial Genetic Tester Adjusted to 1 x 10^{10} cells/ml
(in 100mM PPB pH=7.4)

TX1 Adjusted to 100 mg/ml (in MX⁻ media)

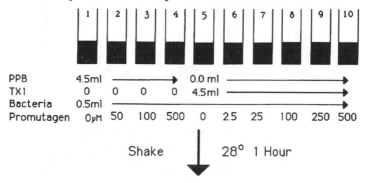

	1	2	3	4	5	6	7	8	9	10
PPB	4.5ml	⟶			0.0 ml	⟶				
TX1	0	0	0	0	4.5ml	⟶				
Bacteria	0.5ml	⟶								
Promutagen	0µM	50	100	500	0	2.5	25	100	250	500

Shake   28° 1 Hour

Place triplicate amounts of 0.5ml of reaction mixture into
2 ml of molten selective medium top agar.

Pour on selective medium plates.

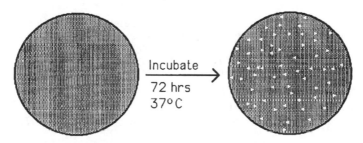

Incubate
72 hrs
37°C

Figure 6.1 Schematic for the plant cell/microbe coincubation assay.

of the promutagen in ≤25 µL dimethyl sulfoxide. Concurrent negative controls consisted of plant and bacterial cells alone, heat-killed plant cells plus bacteria and the promutagen, and both buffer and solvent controls. These components were incubated at 28°C for 1 hr with shaking at 150 rpm. After the treatment time, triplicate 0.5 mL aliquots (~5 × 10^8 bacteria) were removed and added to molten top agar supplemented with 550 µM histidine and biotin. The top agar was poured onto Vogel Bonner (VB) minimal medium plates, incubated for 48–72 hr at 37°C, and revertant his^+ colonies were scored.

Recently, new derivatives of S. typhimurium strain TA98 were developed that possess elevated levels of acetyl-CoA: N-hydroxyarylamine O-acetyltransferase.[17] These new strains include YG1024, which was very sensitive to N-hydroxylated aromatic amines. O-Acetyltransferase acetylates N-hydroxy-lated aromatic amines. This results in the greatly enhanced sensitivity of YG1024

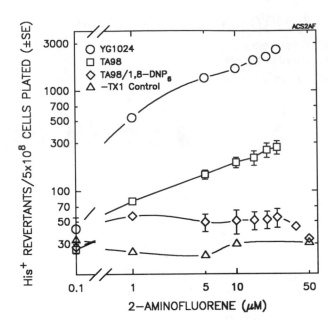

Figure 6.2   Log concentration–response curves illustrating the plant-activation of 2-aminofluorene. For YG1024 and TA98 significant differences ($F_{6, 38} = 750.4, p \leq 0.001$ and $F_{6, 43} = 18.74$, $p \leq 0.001$, respectively) over the negative controls were determined for the concentration range of 1–25 μM. For TA98/1,8-DNP$_6$ no significant difference ($F_{8, 47} = 1.79$) was observed throughout the concentration range of 1–50 μM.

to the plant-activated products of 2-aminofluorene compared to TA98 (Fig. 6.2). An acetyltransferase-deficient strain also exists (TA98/1,8-DNP$_6$).[18,19]

Isolation of Independent Spontaneous *hisD3052* Revertants

For each *S. typhimurium* strain, single colony isolates were collected from master plates. Each isolate was grown overnight in 1 mL of LB medium at 37°C and for each culture, 100 μL was added to molten VB top agar supplemented with 550 μM histidine plus biotin, poured onto a VB plate, and incubated for 72 hr at 37°C. One spontaneous *his*$^+$ revertant was isolated per plate and stored on an agar stab at 4°C. Alternatively each independent revertant was streaked on quartered VB + biotin plates, incubated overnight at 37°C, and stored at 4°C. Five hundred random independent spontaneous mutants were collected for each *S. typhimurium* tester strain.

Isolation of Plant-Activated 2-Aminofluorene-Induced *hisD3052* Revertants

Because of the differential sensitivities of YG1024 and TA98 to plant-activated 2-aminofluorene, both strains were exposed to different concentra-

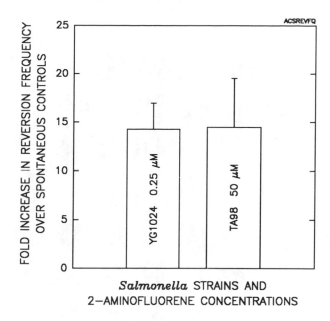

Figure 6.3 A comparison of the increase of plant-activated induced revertants in *S. typhimurium* strains YG1024 and TA98 exposed to 0.25 and 50 μM 2-AF, respectively.

tions of 2-aminofluorene in the coincubation assay to determine the equivalent biological effect. For YC1024, 0.25 μM 2-aminofluorene resulted in an approximate 15-fold increase in mutagenic activity over the spontaneous reversion frequency. For TA98, 50 μM 2-aminofluorene elicited the same fold increase over background (Fig. 6.3). These were the 2-aminofluorene concentrations selected for collecting the plant-activated induced revertants. Induced *his*[+] revertants were collected from three independent experiments. The plant-activated 2-aminofluorene-induced revertants were isolated by picking individual colonies from minimal medium plates and inoculating them in minimal (VB + biotin) liquid medium. These were grown for 1–2 days at 37°C. The revertants were maintained by streaking each revertant on quartered VB + biotin plates, and storing them at 4°C after an overnight incubation at 37°C.

Colony-Probe Hybridization

Each *his*[+] revertant was analyzed by a modified version of the Cebula colony-probe hybridization assay.[9,10] The spontaneous or induced revertants were recovered from their storage plates and grown overnight in VB + biotin liquid medium. Brain heart infusion (BHI) agar plates, a very rich growth medium, were divided into grids of 50 sections and numbered. One drop of each revertant suspension was transferred to its corresponding section. A positive control consisting of a known –2 deletion at D878–885, was spotted to each BHI plate. The BHI plates were made in duplicate and incubated overnight at 37°C. After the colonies grew a disk of No. 541 Whatman filter paper was placed

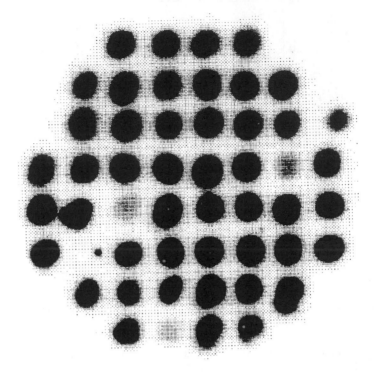

Figure 6.4 A computer scan of an autoradiograph of TC13/TC5-probed TA98 revertants induced
by plant-activated 2-AF.

on the revertant colonies of each BHI plate. The colonies adhered to the filter
disk by applying pressure to the surface of the agar and peeling back the filter.
The disks were placed in a denaturing solution (1.5 M NaCl, 2 M NaOH) for 5
min, irradiated in a microwave oven to lyse the cells, transferred to a neutralizing solution (1 M Tris, 2 M NaCl pH 7.0) for 5 min, and dried overnight.
Filters containing the lysed $hisD3052$ revertants were put in petri plates with
10 mL of hybridization solution (25 mM potassium phosphate buffer, pH 7.4,
5 × SSC, 5 × Denhardt's solution, 100 μg/mL calf thymus DNA, 25%
formamide). An unlabeled competitive probe—TC13 (D892 5'-
GCCGGCAGGCCCTGAGCG-3') was added at a concentration of 30 pmol/
filter to linearize the DNA at a secondary hairpin loop that contains the common
−2 deletion. After 20 min at 60°C, a ^{32}P-labeled probe, TC5 (D874 5'-
CTGCCGCGCCGCGGACACCGC-3') was added at 50 pmol/filter and incubated for 2 hr at 60°C. TC5 is a probe that contains a −2 deletion at D878–885.
The filters were then washed in 3 × SSC for 30 min at 60°C. They were washed
a second time, dried, and exposed to Kodak XAR-5 X-ray film with intensifier
screens overnight at −70°C. The number of revertants that probed for a −2

deletion at D878–885 were recorded. An autoradiogram of a colony-probe hybridization filter is illustrated in Figure 6.4. In this figure three colonies did not hybridize and are presumed to be mutations at some other location in the *hisD3052* region.

PCR and Dot Blot Analysis

The revertant colonies that were negative or unclear in the colony-probe hybridization procedure were grown overnight in LB + biotin medium. The cultures were washed and suspended in 200 µL TE buffer, pH 7.4. Genomic DNA was isolated from the cells using a minipreparation method and the samples were stored at –20°C.[20] A 1:10 dilution of the DNA served as a template to amplify a 635-base pair (bp) DNA fragment containing the *hisD3052* region. Polymerase chain reaction (PCR) amplification was conducted on a DNA thermocycler programmed for 30 cycles of 94°C for 1 min (denaturation), 55°C for 1 min (annealing), and 72°C for 30 sec (extension) using primers AP1 (5′-CGTCTGAAGTACTGGTGATCGCA-3′) and AP3 (5′-CGGGCTAAGTCAGCGACGCTGAG-3′). The PCR products were identified for each sample by agarose gel electrophoresis.

To prepare the dot blots, each well of a 96-well microtiter plate was filled with 195 µL of TE buffer, pH 8.0 (10 mM Tris-Cl, 1 mM EDTA) and a 5-µL aliquot of a specific PCR product. The plate was heated to 95°C for 10 min and the contents from each well were deposited onto a Bio-Rad dot blotter. The vacuum drew each DNA sample onto a charged nylon membrane. The membrane was then rinsed with 500 µL of a 400 mM NaOH solution added to each well. The membrane was removed from the apparatus, rinsed in 30 mL 2 × SSC, and cross-linked with ultraviolet radiation.

Identification of the Hotspot Mutants at D878–885

The Amersham nonradioactive ECL 3′-oligolabeling and detection system was used to identify the –2 hotspot mutation following the general approach of Koch et al.[21] The membrane was probed with TC5 labeled at the 3′-end with a tail of fluorescein-dUTP. The labeled probe was used to detect –2 frameshift mutants at D878–885. The detection of the hybridized probe was based on the protocol furnished by the Amersham kit RPN 2131. The bound labeled TC5 probe was reacted with an antifluorescein horseradish peroxidase conjugate. The reduction of the bound peroxidase is coupled to the oxidation of luminol resulting in the emission of light (λ_{max} 428 nm). The light was detected on X-ray film and serves as the permanent record. The specifics of the probe hybridization are as follows. The membrane was placed in a roller oven and hybridized with 20 pmol of unlabeled TC13 competitive probe in 12 mL hybridization buffer (5 × SSC, 0.1% Amersham hybridization buffer component, 0.02% SDS, 0.5% Amersham blocking agent) at 60°C for 30 min. TC5 labeled probe (158 ng) was hybridized for 1 hr at 60°C after which the mem-

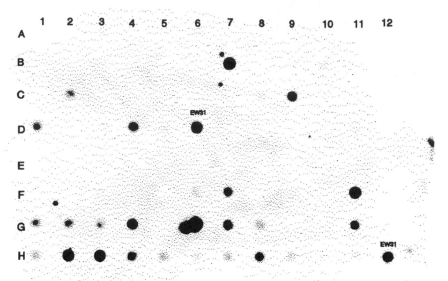

EW YG1024 Plant-Activated 2-AF Induced Mutants (6-2-92)

Figure 6.5   Southern blot analysis of *S. typhimurium* strain YG1024 *hisD3052* revertants (–2 events at D878–885) induced by plant-activated 2-aminofluorene. EW31 was a positive control that was DNA sequenced to confirm the –2 deletion at D878–885.

brane was washed twice in 5 × SSC, 0.1% SDS for 5 min at 40°C. The membrane was washed again twice with 0.75 × SSC, 0.1% SDS for 15 min at 62°C. It was then rinsed in 0.15 *M* NaCl, Tris 0.1 *M*, pH 7.5. The membrane was incubated for 30 min in 20 mL Amersham block solution while shaking at 50 rpm and again rinsed in the above buffer. The membrane was then incubated for 30 min in the Amersham antifluorescein HRP-conjugated 1:1000 dilution in 400 m*M* NaCl, Tris 0.1 *M* pH 7.5, 0.5% bovine serum albumin. After this incubation the same buffer without the albumin was used to rinse the membrane 5× each for 5 min while shaking at 50 rpm. The membrane was treated using the Amersham ECL detection system RPN 2105, wrapped in plastic sheeting, and exposed to X-ray film. The film was developed using standard methods. PCR-amplified genomic DNA that hybridized with probe TC5 contained the –2 deletion at D878–885 (Fig. 6.5) and the DNA from the non-hotspot revertants was saved for sequencing.

RESULTS AND DISCUSSION

Spontaneous Revertants

The acetyltransferase expression was assayed by measuring the acetylation of isoniazid by cell-free extracts. There was a direct relationship in the rate of isoniazid acetylated and the numbers of functional copies of *O*-acetyltransferase

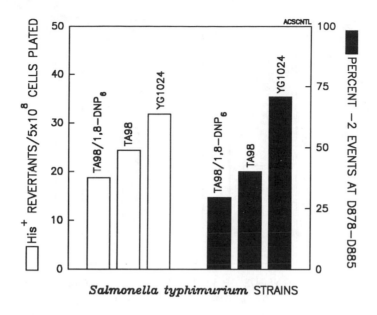

Figure 6.6 The open bars represent the mean spontaneous reversion frequencies for three *S. typhimurium* strains. A significant difference ($F_{2, 281}$ = 37.02, $p \leq 0.001$) was observed in the spontaneous *hisD3052* reversion frequencies for TA98/1,8-DNP₆, TA98, and YG1024. The number of plates assayed for each strain was 92, 98, and 93, respectively. The filled bars represent the percentage of –2 events at D878–885 in spontaneous *hisD3052* revertants for *S. typhimurium* strains TA98/1,8-DNP₆, TA98, and YG1024.

present in the three *S. typhimurium* strains. We observed that the increased expression of *O*-acetyltransferase was correlated with significant increased spontaneous reversion frequencies among the strains (Fig. 6.6).[8] A similar observation was noted by Watanabe et al.[17] *O*-Acetyltransferase is not an essential gene in *S. typhimurium* since TA98/1,8-DNP₆ is viable. However, it appears to interact with some endogenous promutagen such that at higher levels of expression the mean spontaneous reversion frequency was increased 1.7-fold. The mean (±SE) spontaneous reversion frequencies for TA98/1,8-DNP₆ and YG1024 were 18.8 ± 0.7 and 31.8 ± 1.5 revertants per 5×10^8 cells plated, respectively. In addition there was a direct, positive relationship with the increased percentage of –2 events at D878–885 that was concordant with the level of expression of *O*-acetyltransferase (Fig. 6.6). This suggests that the inducer of the increased spontaneous reversion frequency in TA98 and YG1024 is a substrate for *O*-acetyltransferase. We speculate that proteins or amino acids in the medium may be converted to such substrates during the heating process of autoclaving.

Plant-Activated 2-Aminofluorene-Induced Revertants

A comparison of the distribution of –2 events at D878–885 for the spontaneous and plant-activated 2-aminofluorene-induced revertants is presented in Figure 6.7. This –2 deletion accounted for approximately 40% of spontaneous

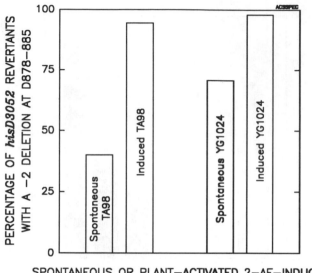

Figure 6.7 Comparison of the spontaneous and plant-activated 2-aminofluorene-induced revertants at the *hisD3052* allele that were identified as a –2 deletion at D878–885.

TA98 revertants. In this study, 1996 plant-activated 2-AF-induced revertants of
TA98 were probed and 1898 or 94% probed as the –2 CG deletion. There was
a significant increase in the frequency of –2 CG deletions in induced revertants
of TA98 as compared to the spontaneous revertants. In strain YG1024, 1616
revertants were probed and 98% of the plant-activated 2-aminofluorene-in-
duced revertants were a result at this deletion. Compared to the YG1024
spontaneous revertants there is a significant increase in this specific –2 deletion.
Cebula and Koch[10] found that 92% of the revertants induced by mammalian S9-
activated acetylaminofluorene were a –2 deletion in this hotspot region.

In conclusion we found that concentrations of plant-activated 2-aminofluorene
that elicited equivalent biological effect (identical mutant fold increases over
spontaneous levels) resulted in a majority of –2 deletions at D878–885 in the
hisD3052 region. The plant-activated 2-aminofluorene-induced spectra were
fundamentally different from the spontaneous spectra. Although further resolution will occur after DNA sequencing, it is evident that exposure to the plant-
activated products of 2-aminofluorene did not result in a high frequency of large
deletions or insertions in either YG1024 or TA98 as compared to the corresponding spontaneous spectra.[12] YG1024 was more sensitive than TA98 to the
plant-activated metabolites of aromatic amines due to elevated levels of *O*-
acetyltransferase. Despite this difference in sensitivity between strains TA98
and YG1024, there is no difference in the frequency of –2 deletions at D878–
885 of these plant-activated 2-aminofluorene-induced revertants.

ACKNOWLEDGMENTS

Research was funded in part by U.S. Air Force Grant AFOSR-91–0432 and by U.S. Environmental Protection Agency Grant R815008. Funds in support of undergraduate research projects were provided by the Institute for Environmental Studies at the University of Illinois at Urbana-Champaign. We acknowledge the kind assistance of Drs. T. A. Cebula, E. Kupchella, and D. M. DeMarini.

REFERENCES

1. Plewa, M. J. Activation of chemicals into mutagens by green plants: A preliminary discussion. *Environ. Health Perspect.* 27, 45, 1978.
2. Plewa, M. J. and Gentile, J. M. Activation of promutagens by plant systems. *Mut. Res.* 197, 173, 1988.
3. Higashi, K. Metabolic activation of environmental chemicals by microsomal enzymes of higher plants. *Mutat. Res.* 197, 273, 1988.
4. Wagner, E. D., Gentile, J. M., and Plewa, M. J. Effect of specific monooxygenase and oxidase inhibitors on the activation of 2-aminofluorene by plant cells. *Mutat. Res.* 216, 163, 1989.
5. Wagner, E. D., Verdier, M. M., and Plewa, M. J. The biochemical mechanisms of the plant activation of promutagenic aromatic amines. *Environ. Mol. Mutagen.* 15, 236, 1990.
6. Plewa, M. J., Smith, S. R., and Wagner, E. D. Diethyldithiocarbamate suppresses the plant activation of aromatic amines by inhibiting tobacco cell peroxidase. *Mutat. Res.* 247, 57, 1991.
7. Plewa, M. J., Weaver, D. L., Blair, L. C., and Gentile, J. M. The activation of 2-aminofluorene by cultured plant cells. *Science* 219, 1427, 1983.
8. Wagner, E. D., Smith, S. R., Xin Hua, and Plewa, M. J. Comparative mutagenicity of plant-activated aromatic amines using *Salmonella* strains with different acetyltransferase activities. *Environ. Mol. Mutagen.* 23, 64, 1994.
9. Cebula, T. A. and Koch, W. H. Analysis of spontaneous and psoralen-induced *Salmonella typhimurium hisG46* revertants by oligodeoxyribonucleotide colony hybridization: use of psoralens to cross-link probes to target sequences. *Mutat. Res.* 229, 79, 1990.
10. Cebula, T. A. and Koch, W. H. Sequence analysis of *Salmonella typhimurium* revertants. *Mutation and the Environment.* Mendelsohn, M. L. and Albertini, R. J., Eds. Wiley-Liss, New York, 1990, p. 367.
11. Bell, D. A., Levine, J. G., and DeMarini, D. M. DNA sequence analysis of revertants of the *hisD3052* allele of *Salmonella typhimurium* TA98 using the polymerase chain reaction and direct sequencing: Application to 1-nitropyrene-induced revertants. *Mutat. Res.* 252, 35, 1991.
12. DeMarini, D. M., Abu-Shakra, A., Bell, D. A., and Levine, J. G. Spectrum of spontaneous frameshift mutations: Analysis of revertants of the *hisD3052* allele of *Salmonella typhimurium* TA98 and TA1538. *Environ. Mol. Mutagen.* 17, Suppl. 19, 21, 1991.

13. Kupchella, E. and Cebula, T. A. Analysis of *Salmonella typhimurium hisD3052* revertants: The use of colony hybridization, PCR, and direct sequencing in mutational analysis. *Environ. Mol. Mutagen.* 18, 224, 1991.

14. Smith, S. R., Wagner, E. D., and Plewa, M. J. Mutational spectra analysis of TA98 and YG1024 revertants induced by the plant activated products of 2-aminofluorene. *Environ. Mol. Mutagen.* 19, Suppl. 20, 58, 1992.

15. Timme, M. J., Shah, A. G., and Plewa, M. J. Mutation spectra of spontaneous mutations at the *hisD3052* allele of *Salmonella typhimurium* strains TA98, TA98/1,8-DNP$_6$ and YG1024. *Environ. Mol. Mutagen.* 19, Suppl. 20, 65, 1992.

16. Plewa, M. J., Wagner, E. D., and Gentile, J. M. The plant cell/microbe coincubation assay for the analysis of plant-activated promutagens. *Mutat. Res.* 197, 207, 1988.

17. Watanabe, M., Ishidate, M. Jr., and Nohmi, T. Sensitive method for the detection of mutagenic nitroarenes and aromatic amines: New derivatives of *Salmonella typhimurium* tester strains possessing elevated *O*-acetyltransferase levels. *Mutat. Res.* 234, 337, 1990.

18. McCoy, E. C., Anders, M., and Rosenkranz, H. S. The basis of the insensitivity of *Salmonella typhimurium* strain TA98/1, 8-DNP$_6$ to the mutagenic action of nitroarenes. *Mutat. Res.* 121, 17, 1983.

19. Saito, K., Yamazoe, Y., Kamataki, T., and Kato, R. Mechanism of activation of proximate mutagens in Ames' tester strains: the acetyl-CoA dependent enzyme in *Salmonella typhimurium* TA98 deficient in TA98/1, 8-DNP$_6$ catalyzed DNA-binding as the cause of mutagenicity. *Biochem. Biophys. Res. Commun.* 116, 141, 1983.

20. Ausubel, F. M., Brent, R., Kingston, R. E., Moore, D. D., Seidman, J. G., Smith, J. A., and Struhl, K. *Current Protocols in Molecular Biology.* Wiley Interscience, New York, 1992.

21. Koch, W. H., Wentz, B. A., and Cebula, T. A. Mutational analysis of *hisG46* reversion using nonisotopic oligonucleotide hybridization analysis. *Environ. Mol. Mutagen.* 19, Suppl. 20, 30, 1992.

CHAPTER 7

Identification of Sulfate-Reducing Bacteria by Hydrogenase Gene Probes and Reverse Sample Genome Probing

G. Voordouw, A. J. Telang, T. R. Jack, J. Foght,
P. M. Fedorak, and D. W. S. Westlake

INTRODUCTION TO SULFATE-REDUCING BACTERIA

Sulfate-reducing bacteria (SRB) are ubiquitously found in anaerobic environments where organic substrates and sulfate are present. Because these bacteria use sulfate as the respiratory substrate (dissimilatory sulfate reduction) they produce copious amounts of hydrogen sulfide. This respiratory endproduct diffuses to the aerobic layers, where it is reoxidized to sulfate by sulfide-oxidizing bacteria. SRB thus form a vital link in the global sulfur cycle.[1] Taxonomic studies based on nutritional requirements and 16S rRNA sequencing have indicated that there are at least 8 different taxonomic groups of the SRB.[2-4] Although these can be catalogued in various ways, e.g., the genus *Desulfotomaculum* is Gram positive, whereas the others are Gram negative, a useful division for understanding the role of SRB in microbial ecology is that between the complete and the incomplete oxidizers.

Complete oxidizers, e.g., *Desulfobacter*, achieve the complete conversion of organic substrates (typically acetate) to CO_2 and H_2S, whereas incomplete oxidizers (e.g., *Desulfovibrio*) convert their organic substrate (e.g., lactate) to acetate, CO_2, and H_2S. SRB do not generally use monomeric or polymeric carbohydrates for their energy metabolism. Instead, they use organic acids (e.g., acetate, lactate, propionate, decanoate) and/or alcohols and/or hydrogen as energy substrates. These are supplied by anaerobic, carbohydrate-degrading, fermentative bacteria such as *Clostridium*. A microbial consortium capable of complete, anaerobic, sulfate-dependent degradation of a polymeric carbohydrate, such as cellulose, to CO_2 and H_2S, thus minimally consists of a fermentative bacterium, an incomplete and a complete sulfate-reducer.

The interest in quantitating the presence of SRB in the environment stems mainly from the fact that the H_2S formed by these bacteria has corrosive

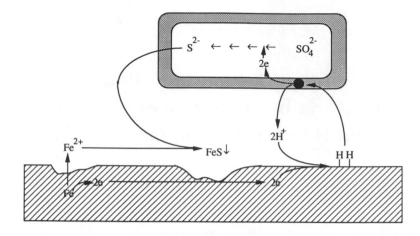

Figure 7.1 Model for corrosion acceleration by SRB. Low potential electrons are abstracted from the metal surface, possibly as hydrogen through a periplasmic hydrogenase, and used to reduce sulfate to sulfide in a cytoplasmic sulfate-reduction pathway in which the enzyme APS reductase functions. Excreted sulfide reacts with iron dissolved in the corrosion process to form insoluble iron sulfide.

properties. In addition, the ability of these bacteria to use hydrogen (or perhaps to abstract low potential electrons directly from a metal surface) is thought to contribute to microbially induced corrosion (MIC) catalyzed by these bacteria.[5] As shown diagrammatically in Figure 7.1, hydrogen consumption and sulfide production could both contribute to the corrosion process, especially in metal pipes subjected to anaerobic conditions. Thus, the potential of SRB to cause MIC is one of the reasons for the interest in quantitating the presence of these bacteria in the environment.

ASSAYS FOR SULFATE-REDUCING BACTERIA

The most probable number (MPN) assay in which a dilution series of a sample is incubated in an anaerobic lactate-sulfate medium is still the most popular test for SRB determination. The formation of black iron sulfide serves as the indicator for the presence of SRB. With commercially available test kits, the amount of work involved in starting an assay is minimal. Problems are primarily related to the fact that (1) long incubation times are often required before the MPN can be read, (2) some genera of SRB (e.g., *Desulfobacter*) are unable to use lactate so the assay is not entirely generic,[2-4] and (3) only a fraction of the bacteria present may be able to grow, following the transfer of a sample from the environment into the test medium.[6]

Two recently designed tests for SRB do not require growth and can thus, in principle, be carried out more rapidly. In the first of these, the activity of the enzyme hydrogenase, considered to be a key enzyme in MIC (Fig. 7.1), is determined as a measure for the presence of SRB. Reduction of redox dyes (e.g., benzyl- or methylviologen) in a hydrogen atmosphere allows colorimetric quantification of hydrogenase activity.[7] In the second test, the presence of adenosine-5′-phosphosulfate (APS) reductase, a key enzyme in the sulfate reduction pathway, is monitored immunologically.[8,9]

Although rapid, these assays are not without their own pitfalls. Microorganisms other than SRB are known to exhibit hydrogenase activity,[10] and reduction of redox dyes can be mediated by a variety of low potential electron donors. Also, APS reductase may not be sufficiently conserved to allow efficient immunological detection in all SRB.[9] In a comparative study, the APS reductase test appeared to give better correlation with MPN determination than the hydrogenase test.[11]

GENE PROBE ASSAYS FOR SULFATE-REDUCING BACTERIA

Hydrogenase and 16S rRNA Gene Probes

In the previous section the need for a generalized SRB assay has been emphasized. Although this feature is important if one is interested only in the total number of SRB, it is unlikely that this information will be useful in elucidating the MIC mechanism, which may require information about the specific SRB and other bacteria present at corrosion sites. Gene probes can rapidly provide the required specificity. The sequencing of 16S rRNA genes has helped to define the phylogenetic relationships of a wide variety of SRB.[3,4] This has also allowed the design of oligonucleotide probes for detection of (1) all eubacteria, or (2) *Desulfovibrio desulfuricans* and *Desulfotomaculum*, or (3) *Desulfovibrio desulfuricans* and *Desulfobacter*.[12] Application of fluorescent oligonucleotide probes, based on 16S rRNA sequences, has allowed individual species to be recognized in a mixed population biofilm.[13]

In our laboratory we have determined the potential of hydrogenase gene probes for the identification of SRB in the environment.[14] SRB of the genus *Desulfovibrio* are known to contain genes for three distinctly different hydrogenases. One set of these genes, encoding the [NiFe] hydrogenase, was detected in each of 22 different *Desulfovibrio* species tested.[14] The labeled [NiFe] hydrogenase genes could thus be used as a *Desulfovibrio*-specific probe, because the genomes of other SRB groups did not hybridize with the probe.[14] The [NiFe] hydrogenase probe was used in conjunction with the Southern blot technique to analyze samples obtained from oil field sites, as shown in Figure 7.2. Twenty samples were obtained from nine different sites in oil fields, from which oil was recovered by water flooding.

The produced oil–water mixture was separated in several production facilities, each consisting of a free water knock out, in which the oil and water were separated, and a water plant, in which the produced water was collected prior to reinjection into the reservoir. A test facility, consisting of five test loops from

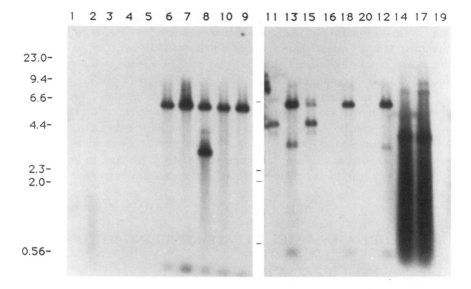

Figure 7.2 Analysis of SRB in oil field samples with a [NiFe] hydrogenase gene probe. SRB in samples listed in Table 7.1 were enriched by growth in 100 mL of Postgate's medium C.[1] DNA extracted from the resulting culture was digested with restriction endonuclease *Eco*RI, electrophoresed through agarose, and blotted. The resulting blot was incubated with a radiolabeled [NiFe] hydrogenase gene probe. Reprinted with permission from *Applied and Environmental Microbiology.*[14]

which samples could be drawn easily, was present at one of the sites. Southern blot analysis allowed *Desulfovibrio* species to be identified in 12 of the 20 samples, while DNA prepared from 8 samples showed no detectable hybridization with the hydrogenase probe (Fig. 7.2).

The positive hybridization indicated the presence of *Desulfovibrio* species in these 12 samples, and comparison of the hybidization patterns suggested the presence of five different *Desulfovibrio* species (Table 7.1). Interestingly, the *Desulfovibrio* species cultured from the free water knock out and water plant were the same at some, but different at other sites (Table 7.1, site V vs. VI and VIII). Some sites (the test loops, as well as VII and IX) appeared to lack culturable *Desulfovibrio*. This does not mean that SRB were completely absent from these sites, because the hydrogenase probe could only detect and distinguish *Desulfovibrio* species and could not be used for identification of SRB from any of the other seven different 16S rRNA groups. The genomes of these other SRB did not hybridize with the [NiFe] hydrogenase probe.[14]

The presence of SRB other than *Desulfovibrio* in any of the 20 samples listed in Table 7.1 thus can not be ruled out. Their presence or absence could, in principle, be established with more generally applicable probes, e.g., 16S rRNA probes that have been used extensively to characterize microbial communities in the environment either directly[13] or following amplification by the

Table 7.1 Presence of *Desulfovibrio* Species Detected by Southern Blotting[a]

Number[b]	Site	Field[c]	Production facility	*Desulfovibrio*[d]
1	I	WW	Test loop line 1	—
2	I	WW	Test loop line 2	—
3	I	WW	Test loop line 3	—
4	I	WW	Test loop line 4	—
5	I	WW	Test loop line 5	—
6	II	WM	Wildmere washtank	D1
7	II	WM	Wildmere water plant 1	D1
8	II	WM	Wildmere water plant 2	D1,D2,(D5)
9	III	WW	Truckpit washtank	D1
10	III	WW	Truckpit water plant	D1
11	IV	WW	Water plant	D3
12	V	WW	Free water knock out	D1,D2
13	V	WW	Water plant	D1,D2
14	VI	WW	Free water knock out	D4
15	VI	WW	Water plant	D1,D3
16	VII	WW	Water plant	—
17	VIII	WW	Free water knock out	D4
18	VIII	WW	Water plant	D1
19	IX	WW	Free water knock out	—
20	IX	WW	Water plant	—

[a] The identification of different species is based on the hybridization pattern with the [NiFe] hydrogenase gene probe, as shown in Figure 7.2.

[b] Numbers correspond to the lanes in Figure 7.2.

[c] Samples were from Wainwright (WW) or Wildmere (WM); see Figure 7.6.

[d] Some samples failed to hybridize (—); D1 is the same as *Lac*3 in Table 7.2.

polymerase chain reaction.[15] Although these probes and the associated methodology have proven their usefulness for detailed characterization of a limited number of samples, we considered the methodology too cumbersome for routine screening of samples.

For analysis of the microbial diversity in large numbers of samples we capitalized on our finding that, due to the large genetic diversity of SRB, the genomic DNA of a given species can be used as a specific probe for itself and closely related species.[16] As an example, specific hybridization only to self or related isolates is shown in Figure 7.3 for the marine SRB *Desulfovibrio salexigens*, the Gram positive *Desulfotomaculum ruminis,* and the thermophilic *"Desulfovibrio" thermophilus*. Genomic DNA hybridization was found suitable for rapid characterization of microbial diversity in the environment, especially when performed in a reverse manner as described in the section below.

Reverse Sample Genome Probing

The data in Figure 7.3 suggest that genomic DNA from SRB is sufficiently dissimilar that samples containing mixed SRB populations taken from from oil fields can in principle be analyzed by probing total DNA extracted from the sample sequentially with genomic DNA probes from different species (i.e., traditional hybridization). If a large population diversity is anticipated, then

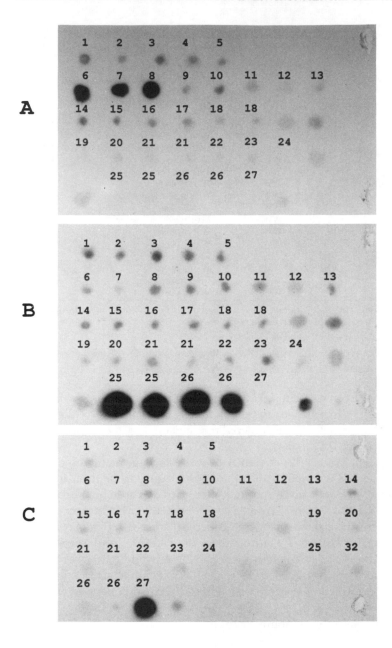

Figure 7.3 Hybridization specificity of genomic DNA probes from SRB. Denatured, chromosomal DNAs from SRB type cultures were spotted on a filter. Each number represents a different type culture.[16] The filters were then incubated with a ^{32}P-labeled genomic DNA probe from (A) *Desulfovibrio salexigens* (spots 6–8 are different strains of *Desulfovibrio salexigens*), (B) *Desulfotomaculum ruminis* (spots 25 and 26 are different strains of *Desulfotomaculum ruminis*), and (C) *"Desulfovibrio" thermophilus* (spot 27 is *"Desulfovibrio" thermophilus*, which has been shown not to belong to the genus *Desulfovibrio*). Reprinted with permission from *Applied and Environmental Microbiology*.[16]

Analysis of total DNA isolated from an oil field or other environmental sample with RSGP using a master filter with 12 standards:

- Isolate total DNA from sample (S).

- Spot denatured DNA (S) on master filter containing denatured DNAs from bacterial standards 1-12.

- Label S to S*. Incubate S* with master filter.

- Develop to see hybridization of S* with master filter.

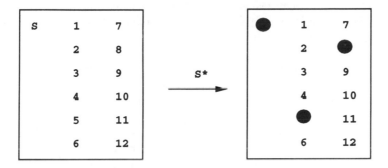

Figure 7.4 Principle of reverse sample genome probing. The master filter contains denatured chromosomal DNAs from 12 standards. Incubation with a labeled sample DNA preparation (S*) indicates the presence of standards 5 and 8 in the sample. Self hybridization of S* with S is the positive control for each RSGP assay. Any number of unique standards can be used in preparing the filter.

conventional genome probing by repeated incubation with different probes is not practical.

Much more rapid analyses can be achieved by carrying out the process in a reverse manner: denatured, genomic DNAs from the species that might be present in the sample are spotted on a filter (the "master filter"), which is then incubated with labeled total DNA extracted from the sample. This technique, reverse sample genome probing (RSGP),[16,17] is outlined in Figure 7.4. Because species of bacteria that are closely related and have strongly cross-hybridizing genomes cannot be discriminated by RSGP, we have introduced the term "standard" to indicate all of these closely related cross-hybridizing groups.[16] Ideally, only standards with little or no cross-hybridization (Figure 7.3) should be included on the master filter. If denatured chromosomal DNAs from *Desulfovibrio salexigens*, *Desulfotomaculum orientis*, and *"Desulfovibrio" thermophilus* are spotted on a filter, then the resulting master filter can be used to determine whether a given sample contains one or more of these three standards.

Examples of RSGP identification of SRB using a master filter with 35 standards are shown in Figure 7.5. The applicability, sensitivity, and accuracy of RSGP identification can be expected to depend on several factors:

1. The master filter formula: if a sample DNA preparation fails to hybridize with the standards on a filter, (but shows self-hybridization in the positive control) then attempts should be made to isolate the standards present in that sample, ideally by colony purification, since the sample apparently contains new, unique standard strains. The master filter formula can then be extended to include these newly isolated standards.
2. A large fraction of bacteria in a sample may not be culturable, underestimating the population diversity at that site.
3. The amount of label incorporated into genomic DNA of a given standard is proportional to its fraction in the mixed population sample DNA (see also "Potential for Quantitative Reverse Sample Genome Probing"). This fraction can be increased by liquid culture enrichment (RSGP with growth) to specifically enhance the signal from certain strains.

The autoradiograms in Figure 7.5 indicate that intense hybridization signals can be obtained by RSGP of DNA preparations from mixed population liquid culture enrichments. However, quantitative information is lost following selection and amplification of bacteria in a sample by growth. Although the data in Figure 7.5 identify SRB standards only qualitatively, the resulting information is nevertheless informative and repeated analysis of samples by RSGP with growth allowed us to distinguish two distinct SRB populations in oil fields in Alberta, Canada.

POPULATIONS OF SULFATE-REDUCING BACTERIA IN OIL FIELDS

RSGP with growth was used for definition of bacterial populations in six different oil fields and one oil storage facility (Fig. 7.6). Samples were taken from the bulk fluid (planktonic) or the metal surface (sessile) and were cultured under conditions outlined in Table 7.2. For culturing, replicate portions of the sample were injected into each of six serum bottles, containing medium according to Pfennig et al.[2] with one of lactate, ethanol, benzoate, decanote, propionate, or acetate as the carbon and energy source.[17]

Following total DNA extraction from each of the six enrichment cultures, RSGP analyses were done at least once and the results were compared to derive the SRB population at each site. The number of RSGP analyses done to deduce SRB belonging to the population at each site is indicated in Table 7.2. In total, 56 sites at the 7 locations were sampled and 367 DNA preparations were analyzed by RSGP to deduce the SRB populations indicated in Table 7.2. Comparison of the data in Table 7.2 has indicated that the population data fall into two clusters,[17] the Wainwright cluster and the Pembina cluster (Fig. 7.7).

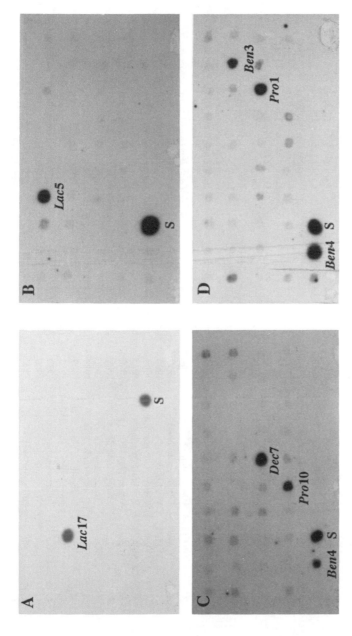

Figure 7.5 Identification of SRB by RSCP following liquid culture enrichment. Denatured chromosomal DNAs from 35 different SRB standards previously isolated from oil fields in Alberta, Canada, were spotted on the master filters.[17] The standards were named with the first three letters of the carbon source used for their first isolation (either lactate, ethanol, benzoate, propionate, decanoate, or acetate) and an identifying number. The master filters were then incubated with the following labeled sample DNA preparations: (A) a colony purified isolate of *Lac17*, (B) a colony purified isolate of *Lac5*, (C) a mixed population liquid enrichment culture of sample PB8 (Table 7.2) which uses decancate as the carbon and energy source, and (D) a mixed population liquid enrichment culture of sample WW5 (Table 7.2) using benzoate as the carbon and energy source. The most prevalent standards reacting with the reverse genome probes in (C) and (D) are indicated. Reprinted with permission from *Applied and Environmental Microbiology*.[17]

Table 7.2 SRB Populations at 56 Sites from Which Samples Were Taken for SRB Enrichment and Subsequent RSGP Analysis

Location[a]	Company	Site	Description	Year[b]	Type[c]	Sal[d]	T(°C)[e]	n[f]	SRB population[g]
Wainwright	A	WW1	Water plant 13	1989	p	s	30	4	Lac4 Lac5 Lac10 Ben1
Wainwright	A	WW2	Water plant 13	1989	s	s	30	3	Lac5
Wainwright	A	WW3	Water plant truckpit	1989	p	s	30	13	Lac4 Lac5 Lac10 Ben1 Ben4 Dec3 Pro12
Wainwright	A	WW4	Water plant	1989	s	s	30	5	Lac4 Lac5 Lac10 Ben1 Ben4 Pro12
Wainwright	A	WW5	Truckpit	1989	p	s	30	10	Lac4 Lac5 Lac10 Ben1 Ben4 Dec3 Pro1 Pro12
Wainwright	A	WW6	Water plant 1	1989	s	s	30	3	Lac6 Eth3 Ben1
Wainwright	A	WW7	Water plant 20	1989	p	s	30	7	Lac4 Lac5 Lac10 Ben1 Pro1 Ace1
Wainwright	A	WW8	Water plant 28	1989	s	s	30	4	Lac4 Lac6 Lac10
Wildmere	A	WM9	Washtank	1989	s	s	50	0	No SRB recovered
Wildmere	A	WM10	Lower water plant	1989	s	s	35	12	Lac4 Lac5 Lac10 Ben1 Dec8 Pro1
Wainwright	A	WW11	Water plant 20	1990	p	s	30	5	Lac4 Lac6 Eth3 Ben1 Dec1 Pro1
Wainwright	A	WW12	Water plant 20	1990	s	s	30	6	Lac6 Eth3 Ben1 Pro4 Ace1 Ace3
Wainwright	A	WW13	Truckpit	1990	p	s	30	7	Lac6 Pro4 Ace1 Ace3 Ace4
Wainwright	A	WW14	Water plant truckpit	1990	p	s	30	7	Lac6 Ben1 Pro4 Ace1 Ace4
Wainwright	A	WW15	Water plant 13	1990	p	s	30	11	Lac6 Ben1 Pro4 Ace1 Ace3
Wainwright	A	WW16	Water plant 13	1990	s	s	30	7	Lac6 Eth3 Ben1 Dec1 Dec3 Pro4
Wainwright	A	WW17	Water plant 1	1990	p	s	30	8	Lac4 Lac6 Lac12 Lac21 Eth3 Ben1 Ben3 Ace1 Ace3
Wainwright	A	WW18	Water plant 28	1990	p	s	30	10	Lac4 Lac6 Lac12 Lac21 Ben1 Dec8 Pro7 Pro11 Pro12
Wainwright	A	WW19	Water plant 6	1990	p	s	30	6	Lac6 Ben1 Dec1 Ace1 Ace3
Wildmere	A	WM20	Lower water plant	1990	s	b	22	4	Lac10 Ben1 Ben4
Wildmere	A	WM21	Washtank	1990	s	b	22	5	Lac4 Lac5 Ben4 Dec1
Wainwright	A	WW22	Water plant truckpit	1990	s	b	22	6	Lac3 Lac6 Ben1 Pro1
Wainwright	A	WW23	Washtank truckpit	1990	s	b	22	6	Lac3 Lac6 Ben1 Ben4 Dec1
Wainwright	A	WW24	Water plant 20	1990	s	f	22	3	Lac3 Lac6 Ben1 Ben4
Edmonton	B	ED1	Storage tank 21	1989	p	f	22	3	Lac8 Lac15
Edmonton	B	ED2	Storage tank 22	1989	p	f	22	4	Lac1,2 Lac8 Lac15
Edmonton	B	ED3	Storage tank 23	1989	p	f	22	2	Dec3
Virginia Hills	C	VH1	Induced gas flow inlet	1989	p	s	22	1	Lac6
Virginia Hills	C	VH2	Production water dump	1989	p	s	22	10	Lac5 Lac6 Ben1 Dec1 Dec3
Virginia Hills	C	VH3	Induced gas flow outlet	1989	p	s	22	1	No confirmed identification
Virginia Hills	C	VH4	Free water knock out	1990	p	s	50	0	No SRB recovered
Virginia Hills	C	VH5	Induced gas flow outlet	1990	p	s	22	7	Lac6 Lac21 Eth3 Ben1 Dec8 Ace1
Virginia Hills	C	VH6	Sandfilter outlet	1990	p	s	22	3	Lac6 Lac12

Location[a]	Sample		Source	Year[b]	Type[c]	Salinity[d]	Temp[e]	No.[f]	SRB population[g]
House Mountain	HM7	C	Unfiltered water	1989	p	s	22	6	Lac6 Ben1 Dec3
House Mountain	HM8	C	Pipe scrapings	1989	p	s	22	3	Lac6 Ben1 Dec3 Ace4
House Mountain	HM9	C	Unfiltered water	1989	p	s	22	3	Lac6 Ben1 Dec3
House Mountain	HM10	C	Produced well water	1990	p	s	22	8	Lac6 Ben1 Dec1 Dec3 Pro4 Ace4
House Mountain	HM11	C	Unfiltered combined water	1990	p	s	22	7	Lac6 Ben1 Dec1 Dec3 Pro4 Pro12 Ace1
House Mountain	HM12	C	Filtered combined water	1990	p	s	22	6	Lac6 Dec3 Pro4
Harmattan	HR13	C	Produced water	1988	p	b	35	13	Lac3 Lac4 Lac10 Ben3 Ben4 Dec1 Pro4
Harmattan	HR14	C	Flow splitter	1988	p	b	35	11	Lac6 Lac12 Lac21 Eth2 Ben4
Harmattan	HR15	C	Well 11-6	1988	p	b	35	5	Lac4 Lac6 Lac12 Lac15 Lac21 Ben4
Harmattan	HR16	C	Produced water	1990	p	b	35	1	No confirmed identification
Harmattan	HR17	C	Flow splitter	1990	p	b	35	3	Lac21
Pembina	PB1	D	Easyford battery	1990	p	f	22	8	Lac12 Lac15 Pro5
Pembina	PB2	D	Easyford battery	1990	p	f	22	6	Lac12 Dec4 Pro5
Pembina	PB3	D	Water 2 stage separator	1990	p	f	22	6	Lac12 Lac15 Pro5
Pembina	PB4	D	Bear Lake, H battery	1990	p	f	22	10	Lac3 Lac12 Lac15 Dec4 Pro5
Pembina	PB5	D	NW Pembina B battery	1990	p	f	22	5	Lac15 Pro5
Pembina	PB6	D	Winfield A battery	1990	p	f	22	7	Lac3 Lac7 Lac12
Pembina	PB7	E	Injection water	1990	p	f	22	11	Lac12 Lac15 Dec4 Pro10
Pembina	PB8	E	8-2 skimmer	1990	p	f	22	11	Lac12 Lac17 Eth2 Ben4 Ben6 Dec7 Pro10 Ace5
Pembina	PB9	E	8-2 treater	1990	p	f	22	5	Lac1,2 Lac12 Eth2 Ben6 Pro5 Pro10 Ace5
Pembina	PB10	E	Injection well 6-1	1990	p	f	22	16	Lac12 Lac15 Lac17 Ben4 Ben6 Dec6 Pro10 Ace5
Pembina	PB11	E	Reservoir	1990	p	f	22	15	Lac1,2 Lac12 Lac15 Lac21 Dec4 Pro5
Pembina	PB12	E	16-7 produced water	1990	p	f	22	14	Lac12 Lac15 Dec4 Dec6 Pro5 Ace5

[a] Locations where samples were taken. All represent oil fields, except Edmonton, which represents an oil storage facility.

[b] Year in which the sample was collected.

[c] Sample type, either planktonic (p) or sessile (s), see text.

[d] Salinity used for cultivation, either saline (s), brackish (b), or fresh water (f).

[e] Temperature used for cultivation of the sample.

[f] Number of RSGP assays performed to arrive at the SRB population.

[g] SRB population as determined by RSGP assays.

Figure 7.6 Locations of oil fields and an oil storage facility (ED) in Alberta, Canada from which samples were obtained for analysis by RSGP with growth. The map locations of Calgary (C) and Edmonton (E), separated by 277 km, are indicated. Fields were Harmattan (HR), House Mountain (HM), Pembina (PB), Virginia Hills (VH), Wildmere (WM), and Wainwright (WW). Reprinted with permission from *Applied and Environmental Microbiology*.[17]

The Wainwright cluster comprises all of the sites from Harmattan, House Mountain, Virginia Hills, Wildmere, and Wainwright, whereas the Pembina cluster comprises only the sites from the Pembina field.

The large difference in the two SRB populations can be appreciated from Figure 7.8, where each confirmed observation of an SRB standard has been plotted for the Wainwright cluster and the Pembina cluster. Of 34 standards, 18 were unique to the Wainwright and 10 were unique to the Pembina cluster, while only 6 standards were common to both environment types.

The discriminating factor between the two environments is the salt concentration of the oil field production waters. Those in the Wainwright cluster are saline (10–50 g of NaCl/L), while those in the Pembina cluster are fresh water (less than 2 g of NaCl/L). Work by Cord-Ruwisch et al.,[18] in which numbers of different types of SRB were evaluated by colony counts, has also indicated that the salinity of the production waters is an important determinant of the SRB population.

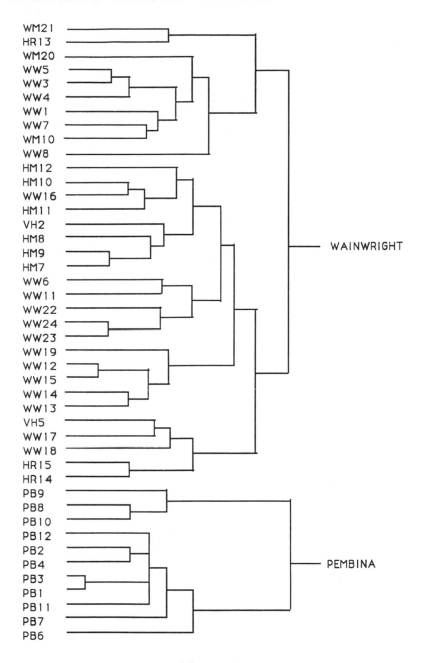

Figure 7.7 Comparison of SRB population data derived by RSGP with growth for 44 of the 56 sites listed in Table 7.2. Only sites for which more than three SRB standards were identified were used for the comparison. Reprinted with permission from *Applied and Environmental Microbiology*.[17]

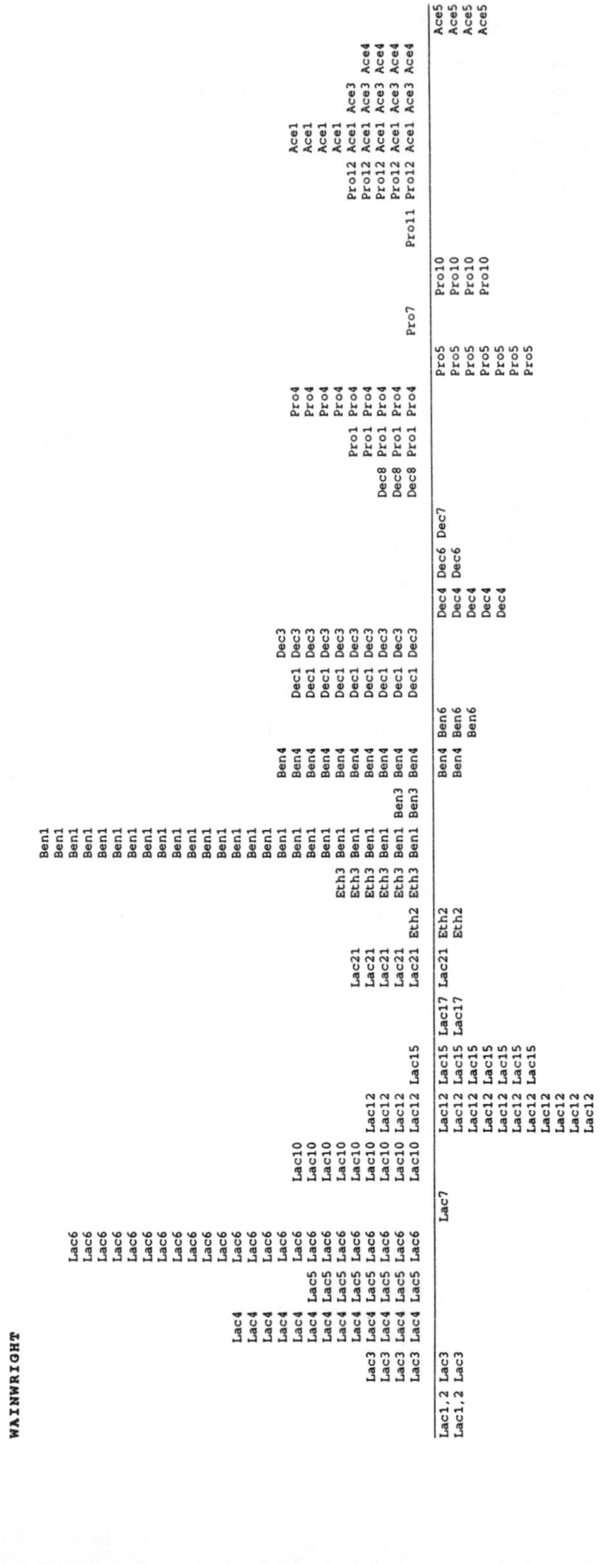

WAINWRIGHT

PEMBINA

Figure 7.8 Frequency of observation of SRB standards in samples from the Wainwright cluster (above the line) and the Pembina cluster (below the line).

POTENTIAL FOR QUANTITATIVE REVERSE SAMPLE GENOME PROBING

Although RSGP with growth is useful to define the SRB standards that are present at a field site, the data obtained have only qualitative significance. For quantitative analysis the growth step must be eliminated, since growth selectively shifts the relative numbers of bacteria from those initially present in the field towards those strains capable of growing under the conditions provided. The problem in RSGP analysis without growth is that the SRB standards present on the filter may represent only a small proportion of the total DNA isolated from an environmental sample, which also will contain non-SRB microbes. Upon labeling of a DNA preparation isolated from such a sample, a proportionally small fraction of the added label will be incorporated into the genomic DNA of the SRB present.

If the resulting reverse genome probe is subsequently hybridized with a master filter, weak signals are expected. Dot blot studies in which either the amount (c_x) of a given standard DNA spotted on a filter, or the fraction of label (f_x) incorporated into standard x during the labeling reaction were varied, have shown that the hybridization intensity I_x for a given standard x is

$$I_x = k_x f_x c_x \qquad (7.1)$$

where the rate constant k_x is inversely proportional to genome complexity.[19] Labeling of the extracted total DNA in the presence of a known amount (f_s) of an added internal standard and with a known amount (c_s) spotted on the filter allows calculation of the fraction f_x of standard x in the sample as:

$$f_x \propto (I_x/I_s) \times (k_s/k_x) \times (c_s/c_x) \times f_s \qquad (7.2)$$

Importantly, sensitivity (for determination of f_x from the measured I_x and I_s) can be increased by applying more denatured standard DNA to the master filter (increasing c_x), since I_x may be expected to increase proportionally. Equation 7.2 has been used for analysis of DNAs isolated directly from production water samples and corrosion coupons.[19]

CONCLUSIONS

Although a variety of methods have been developed to analyze the presence of bacteria in the environment with the use of gene probes, RSGP is exceptional in its ability to characterize the diversity of bacteria present at a given environmental site. Analysis of samples from sites in six oil fields in Alberta by RSGP with growth has indicated that 34 different SRB are frequently cultured from these fields, although comparison of the derived SRB populations has shown that there are two distinctly different SRB communities depending on the salinity of the oil field production waters. Finally, RSGP without growth can be used to quantitate the microbial population at oil field sites, provided high sensitivity master filters are used for the analysis.

ACKNOWLEDGMENTS

This work was supported by a Strategic Grant of the Natural Science and Engineering Research Council of Canada (NSERC) and by an NSERC Collaborative Research and Development Grant with Novacor Research and Technology Corporation.

REFERENCES

1. Postgate, J. R. *The Sulphate-Reducing Bacteria*, 2nd ed. Cambridge University Press, Cambridge, 1984.
2. Pfennig, N., Widdel, F., and Trüper, H. G. Dissimilatory sulphate-reducing bacteria. *The Prokaryotes*. Starr, M. P., Stolp, H., Trüper, H. G., Balows, A., and Schlegel , H. G., Eds. Springer-Verlag, New York, Vol. 1, 1991, p. 926.
3. Devereux, R., Delaney, M., Widdel, F., and Stahl, D. A. Natural relationships among sulfate-reducing eubacteria. *J. Bacteriol.* 171, 6689, 1989.
4. Devereux, R., He, S.-H., Doyle, C. L., Orkland, S., Stahl, D. A., LeGall, J., and Whitman, W. B. Diversity and origin of *Desulfovibrio* species: phylogenetic definition of a family. *J. Bacteriol.* 172, 1990, 3609.
5. Hamilton, W. A. Sulphate-reducing bacteria and anaerobic corrosion. *Annu. Rev. Microbiol.* 39, 195, 1985.
6. Byrd, J. J., Leahy, J. G., and Colwell, R. R. Determination of plasmid DNA concentration maintained by nonculturable *Escherichia coli* in marine microcosms. *Appl. Environ. Microbiol.* 58, 2266, 1992.
7. Bryant, R. D., Jansen, W., Boivin, J., Laishley, E. J., and Costerton, J. W. Effect of hydrogenase and mixed sulfate-bacterial populations on the corrosion of steel. *Appl. Environ. Microbiol.* 57, 2804, 1991.
8. Odom, J. M., Gawell, L. J., and Ng, T. K. Sulfate-reducing bacteria detection, DuPont Company, Patent Application ICR 7780-USSN946, 547.
9. Odom, J. M., Jessie, K., Knodel, E., and Emptage, M. Immunological crossreactivities of adenosine-5'-phosphosulfate reductases from sulfate-reducing and sulfide-oxidizing bacteria. *Appl. Environ. Microbiol.* 57, 727, 1991.
10. Voordouw, G. Evolution of hydrogenase genes. *Adv. Inorg. Chem.* 38, 397, 1992.
11. Little, B. Progress in MIC testing. Abstract and lecture at the International Symposium on Microbiologically Influenced Corrosion (MIC) Testing, Sponsored by ASTM committee G-1 on Corrosion of Metals, November 1992, Miami.
12. AccuSearch. Chemiluminescent DNA probe and detection system. Sulfate-reducing bacteria. *Gen-Probe^R* 1990.
13. Amann, R. I., Stromley, J., Devereux, R., Key, R., and Stahl, D. A. Molecular and microscopic identification of sulfate-reducing bacteria in multispecies biofilms. *Appl. Environ. Microbiol.* 58, 614, 1992.

14. Voordouw, G., Niviere, V., Ferris, F. G., Fedorak, P. M., and Westlake, D. W. S. The distribution of hydrogenase genes in *Desulfovibrio* and their use in identification of species from the oil field environment. *Appl. Environ. Microbiol.* 56, 3748, 1990.

15. Giovannoni, S. J., Britschgi, T. B., Moyer, C. L., and Field, K. G. Genetic diversity in Sargasso Sea bacterioplankton. *Nature (London)* 345, 60, 1990.

16. Voordouw, G., Voordouw, J. K., Karkhoff-Schweizer, R. R., Fedorak, P. M., and Westlake, D. W. S. Reverse sample genome probing, a new technique for identification of bacteria in environmental samples by DNA hybridization, and its application to the identification of sulfate-reducing bacteria in oil field samples. *Appl. Environ. Microbiol.* 57, 3070, 1991.

17. Voordouw, G., Voordouw, J. K., Jack, T. R., Foght, J. M., Fedorak, P. M., and Westlake, D. W. S. Identification of distinct communities of sulfate-reducing bacteria in oil fields by reverse sample genome probing. *Appl. Environ. Microbiol.* 58, 3542, 1992.

18. Cord-Ruwisch, R., Kleinitz, W., and Widdel, F. Sulfate-reducing bacteria and their activities in oil production. *J. Petrol. Technol.* 39, 97, 1987.

19. Voordouw, G., Shen, Y., Harrington, C. S., Telang, A. J., Jack, T. R., and Westlake, D. W. S. Quantitative reverse sample genome probing and its application to oil field production waters. *Appl. Environ. Microbiol.* 59, 4101, 1993.

CHAPTER 8

The Uses and Abuses of Immunological Testing

Dennis K. Flaherty

INTRODUCTION

A biomarker is a biochemical or a cellular endpoint used to measure a change in a biological system as a consequence of exposure to a specific xenobiotic or toxic material. The definition can be expanded to include biological changes induced by direct or indirect effects of exposure to toxic materials. Some biomarkers may also be used as a reflection of a health-related effect. Biomarkers should meet minimum criteria. They should be chemical specific, detectable in trace quantities, relatable to past exposure, and obtained by noninvasive techniques.[1]

Since the immune system normally responds to any foreign material, it is conceivable that certain immune endpoints such as the presence of chemical specific antibodies in the blood can be used as biomarkers of exposure. It is also possible that exposure to chemicals alters the number or function of immunocompetent cells in human peripheral blood. It has been suggested that lymphocyte phenotyping by flow cytometry or data from *in vitro* assays could be used as indices of exposure. This chapter will briefly describe several of these immune response endpoints and how they may be appropriately used as possible biomarkers of exposure.

CHEMICAL-SPECIFIC ANTIBODIES AS AN INDEX OF EXPOSURE

To elicit an antibody response, reactive chemicals may bind to a number of host proteins such as albumin or hemoglobin. These chemical–protein conjugates interact with the cells in the immune system which produce antibodies directed toward the chemical. Immunogenic chemicals must meet certain criteria. Small molecular weight chemicals must have a defined stereochemistry due to aromatic rings and/or double bonds. Additionally, the chemical must be able to bind to protein under conditions of normal pH and body temperature in a saline solution. Hence, chemicals such as trimellitic anhydride would elicit antibody production but maleic acid would not be immunogenic. Also, chemical–protein conjugates must persist in the body for 10–14 days so that the cells of the immune system have ample time to recognize the foreign material and produce antibodies.[2]

Although the concept of antigen–specific antibodies as biomarkers has not been fully explored, there is evidence suggesting that some persons exposed to chemicals produce antibodies directed toward the specific chemical. Antibodies directed toward toluene diisothiocyanate (TDI), phthalic anhydride, methyl isocyanate, and formaldehyde have been reported in the sera exposed of subjects.[3-6]

Antibodies Elicited Following Xenobiotic Exposure

Several types of xenobiotic or chemical-specific antibodies can be elicited following exposure. The IgM class of antibody is formed early in the immune response, usually after an initial exposure to foreign material. After repeated exposures, a different class of antibody (IgG) is formed. The IgG isotype differs from IgM in both molecular weight and overall structure. In some instances, the determination of the class of antigen-specific antibody may have some forensic significance in the documentation of single or repeated exposures.

The presence of antigen specific IgM or IgG in peripheral blood only signifies a normal immunological response to foreign material. A similar response would occur after exposure to infectious agents (e.g., influenza) or innocuous materials (e.g., pollens). Therefore, no pathological significance should be ascribed to the presence of xenobiotic-specific IgG and IgM antibodies in peripheral blood.

A Common Method Used to Detect Chemical-Specific Antibody

The method most commonly used to detect and quantify antigen-specific immunoglobulin is the solid phase enzyme-linked immunosorbent assay (ELISA). In the assay, test sera are incubated with solid phase immobilized large (3000 MW) or small molecular weight chemical linked to proteins. After washing to remove unbound antibody, a heterologous antibody directed to the human immunoglobulin is added to the solid phase antigen. The heterologous antibody is unique in that alkaline phosphatase is covalently bound to the molecule. Following washing to remove the unbound second antibodies, enzyme substrate (p-nitrophenol phosphate) is added to the tube. Because the only variable in the assay is the antibody in the test serum, the amount of p-nitrophenol formed is directly proportional to the initial concentration of antibody in the test sera. The ELISA has several advantages. Besides being inexpensive, the assay is sensitive in the nanomole range. Moreover, large numbers of samples can be processed in a short period of time.

The quantity of specific antibody in serum is usually determined by diluting the serum to a point where antibody reactivity is no longer detectable. This laboratory endpoint is termed the "titer" of the test sample. Antibody concentrations in a strong positive sera may encompass up to six log^{10} dilutions.[7]

Weakly positive serum samples can be differentiated from negative samples by statistical means. Because the data are often distributed with a positive skew, nonparametric statistical methods can be used in the segregation process. If a large number of negative sera are available, the upper limit of normal can be set at the 97.5 or the 100th percentile. It is also possible to calculate the confidence interval or variance ratios.[8,9]

Some critics of immunoassays have suggested that, because of the normal metabolism of chemicals, it is difficult to define whether the parental compound or one of the metabolites should be used in immunoassays to detect the presence of antibodies. Metabolism becomes a concern only if the degradation alters the portion of the molecule reactive with the antibody (e.g., aromatic ring). In some instances, the parental compound is used in the assay or it may be useful to determine antibodies directed toward a sentinel, long lasting metabolite.

The utility of using antibodies as an index of exposure is limited by the half-life of the antibody. The half-life of IgG and IgM in peripheral blood is approximately 23 and 5 days, respectively.[10] Because there are high levels of IgG (800–1600 mg/dL) or IgM (40–120 mg/dL) in serum, antibodies may persist for extended periods of time. After a single exposure, antigen-specific antibody levels can persist for 6 months. Antibodies may persist for 2 to 5 years after repeated exposures.

Specific Antibody as an Index of a Health-Related Effect

In blood samples, the presence of chemical-specific antibody of the IgE isotype only suggests the possibility of an immediate allergic reaction (e.g., asthma or urticaria). Because antigen-specific IgE is found in normal and nonsymptomatic subjects, the correlation between the presence of antigen-specific IgE in serum and allergic symptoms may be only 70%.[11] The lack of a correlation may be due to the fact that basophils or mast cells from normal or nonsymptomatic subjects have low numbers of receptors for IgE or that the receptors are occupied with IgE directed toward other allergens.

A Method Used to Detect Allergic Antibody

The presence of chemical specific IgE can be determined in the radioallergosorbent test (RAST) or conventional ELISAs.[11] The major difference between the two assays is that the RAST uses a radiolabeled antihuman IgE whereas the indirect ELISA uses an antihuman antibody–enzyme complex.

Table 8.1 Common Lymphocyte Phenotypes Determined by Flow Cytometry

Total T cell population	CD3
Helper/amplifier T cells	CD4
Cytotoxic/suppressor T cells	CD8
B cells	CD19
NK cells	CD16

Because of the problems with interpretation of the RAST assays, the low concentration in serum (0.3 mg/dL), and the short serum half-life (2.5 days) of IgE, other tests can be used as an adjunct to the clinical diagnosis of immediate allergic reactions. The histamine release assay, not commonly available in most laboratories, measures the release of histamine from IgE-sensitized basophils. For unknown reasons, IgE on the surface of basophils is protected from normal catabolism. Thus, the cell bound IgE is present for extended time periods. Since histamine is the major pharmacological mediator contracting smooth muscle in the lung and increasing vascular permeability of vessels in the skin, there is a better correlation between IgE-allergen-induced histamine release from sensitized basophils and allergic symptoms.

ALTERATIONS IN LYMPHOCYTE SUBSETS AS AN INDEX OF EXPOSURE

New technology has evolved that permits the rapid identification and enumeration of lymphocyte subsets in the blood. Using a flow cytometer, peripheral blood lymphocytes are placed in a high speed laminar capillary flow system. Laser assisted technology measures the size and granularity and any fluorescence associated with each cell passing the laser and fluorescence detectors. Cells are counted in a rapid manner, usually 30,000–90,000 cells/min.[12] In immunological applications, functional subsets of lymphocytes (T, B, and natural killer cells) can be identified via the use of monoclonal antibodies directed toward epitypic surface markers on lymphocyte subsets (Table 8.1). In the traditional medical setting flow cytometry technology has been used to provide mechanistic information that confirms an already tentative medical diagnosis. In a digression from the traditional approach, one might suggest that alterations in peripheral blood subsets, as determined by flow cytometry, in normal healthy subjects can be used as a biomarker of exposure or possible health related effects after exposure to xenobiotics.

Although the technology is attractive and the concept sound, data from studies correlating exposure and effects with changes in peripheral blood lymphocytes are difficult to interpret because of scientific and technical problems associated with the studies. Some of the most common failings of such studies are the lack of proper experimental design, the lack of proper control populations, failure to recognize technical, internal, and external biological factors influencing the data from flow cytometric studies, and improper statistical analyses and presentation of data.

Proper Experimental Design

Initially, a careful health history of each test subject must be determined. The history should determine the presence of preexisting conditions (e.g., diseases of the immune system, infections, medications) influencing the distribution of lymphocyte subsets. To minimize the effects of biological rhythms, blood should be drawn from all test subjects at the same time of day and placed in the proper anticoagulant or properly prepared for shipment. An age- and sex-matched control group (with lifestyles and smoking habits similar to the test subjects) from the same geographic area should be included in the study. In this scenario, blood from both the test group and the control subjects should be drawn, transported, and analyzed in a similar manner. Besides negating data comparisons between blood analyzed immediately (normal range in most laboratories) and transported blood, aberrations in the lymphocytes from the control population would be apparent.

Because of the numerous external and internal factors influencing the distribution of lymphocytes in peripheral blood, an aberrant observation of a lymphocyte subset at a single point in time is difficult to interpret. To ensure that the observation is reproducible, all aberrant findings should be confirmed in additional flow cytometry studies, using the same panel of monoclonal antibodies, at several points in time.

Technical Factors Influencing Lymphocyte Subsets

Technical factors involved in sample preparation, shipping, and storage may also influence the data obtained from flow cytometry studies. Since flow cytometry determinations utilize viable, single cell preparations, care must be taken to prevent the clotting of the test serum sample used. Heparin and disodium EDTA are the anticoagulants of choice. Dipotassium EDTA is often used as the anticoagulant when samples are analyzed within 4 hr of collection. After 4 hr, this anticoagulant fails to maintain the viability and morphology leukocytes necessary for cytometric analyses.[13,14] Conversely, blood collected in heparin remains stable for 24 hr. Hence, heparin is used as the anticoagulant of choice when blood samples are obtained for shipment to distant laboratories for flow cytometric analyses. However, additional steps must be undertaken to preserve the morphology of cellular subsets.[15] The best method of transporting blood involves isolation of the buffy cell coat and dilution in tissue culture media (e.g., RPMI-1640) containing low concentrations of serum. Even under optimal conditions, storage of blood may increase the number of CD4 cells in the test sample.[16]

Because of technical factors, it is often difficult to enumerate select or aberrant subsets in peripheral blood. One factor is the cross-reactivity of monoclonal antibodies used in some assays. For example, CD24 antibodies react with a minor population of B lymphocytes (1%) and granulocytes. Hence, small numbers of granulocytes in the test lymphocyte population would yield a false-positive reaction in single color phenotyping. Although rarely implemented by investigators, the possibility of false-positive reactions can be eliminated by dual color phenotyping which clearly defines the presence of CD24 markers only on B lymphocytes. Also, there are other factors that reduce the ability of the flow cytometer to detect lymphocyte subsets. Some phenotypic markers are expressed in low numbers on the surface of the cell. Additionally, the antibody used to identify the marker may bind weakly to the epitope. In both cases, the cells would yield only weak fluorescence, making it difficult to distinguish between stained and unstained cells.

External and Internal Factors Influencing Lymphocyte Subsets

Most investigators fail to consider the fact that the immune system is dynamic in nature responding to external xenobiotic stimuli. For example, infections may increase or activate select lymphocyte populations increasing the numbers in peripheral blood. Conversely, low numbers of immunocompetent cells may be observed because the cells marginate or localize in areas of infection. Other internal factors can also alter the distribution of lymphocytes in peripheral blood.

Biological rhythms of various types exert influences on lymphocyte subsets. Because of ultraradian (less than 24 hr) rhythms, there may be lower total numbers of lymphocytes in the peripheral blood at several time points during a 24-hr day and, within the lymphocyte population, CD4 cells may be decreased by as much as 50%.[17-19] Although not clearly defined, 7 (ciraseptan) and, possibly, 28 day biorhythms may also affect lymphocyte population dynamics in circulating blood.[17]

The numbers of lymphocytes and constituent subsets in blood vary according to the age of the test subject. Both the percentage and absolute number of CD3 cells per mm^3 are decreased in aged men and/or women.[20] Besides the decreases in total T cell numbers, subjects also have decreased numbers of lymphocytes (CD4, CD45R), which amplify immunosuppression.[21] The relationship between the loss of the CD4, CD45R lymphocyte, and the onset of autoimmune diseases of the elderly is unclear.

The distribution of lymphocyte subsets in peripheral blood can be altered by exercise. Exercise decreases the number of CD3 and CD4 cells in peripheral blood while increasing the percentage of CD16 (natural killer) cells. After cessation of vigorous exercise, the number of CD4 cells returns to normal within 120 min, whereas the NK cell numbers return to baseline only after 24 hr.[22] Other data suggest that the number of CD8 cells is increased by vigorous physical activity.

Prescription drugs can also cause shifts in subpopulations of peripheral blood lymphocytes. Cimetidine (a common medication used to reduce gastric acid secretion) and epinephrine increase the number of CD4 and/or CD8 cells in the blood.[23] Conversely, corticosteroids alter lymphocyte trafficking in the body lowering the number of CD4 cells in the peripheral blood. The number of antibody-producing B cells can also be altered by drugs. Following the administration of insulin, there is an increase in the number of B cells in the blood.[24]

The most common external factor influencing the distribution of lympho-cyte subpopulations is a virus infection.[25] Different viruses evoke disparate immunological responses. Infection with influenza, cytomegalovirus or Epstein–Barr viruses evoke a T cytotoxic cell response.[26] These cells lyse infected target cells expressing MHC I markers and viral-induced antigen. Natural killer cells lyse non-MHC expressing target cells infected with herpes, adenovirus, and cytomegalovirus.[27-32] Since some viral infections are spread by close contact for extended periods of time, the effects of clinical or even subclinical infections on the distribution of lymphocytes may be more pronounced in family units. Therefore, one should be cautious when interpreting altered CD8 or CD16 numbers from test subjects potentially exposed to xenobiotics or other environ-mental agents.

Light to heavy cigarette smoking can also alter both the total white blood cell count and the percentage of lymphocytes. Among the subsets, lower percentages of CD4 cells with increased CD8 cells have been reported.[33] Other studies have suggested that the number of NK cells is reduced in the blood of smokers.[34]

Proper Expression of Experimental Data

It is an abuse of the technology to present lymphocyte phenotyping data solely as a percentage of the total lymphocyte population. This method of presentation assumes that the total white blood cell (WBC) population and the percentage of lymphocytes within the total white cell population remain constant. In fact, the total WBC, the lymphocyte population, and the subsets within the lymphocyte population fluctuate independently. Because of these multivariate interactions, flow cytometry data are best expressed in absolute numbers of cells/mm^3. This type of data expression yields more biologically relevant information.

Statistical Interpretation of Data

Significant differences between the test and control populations are usually determined relative to the normal expected range or in terms of probability measurements. In most institutions, the normal range for immune parameters is defined as the mean value plus or minus two standard deviations. In essence, the normal range states that 95 of 100 persons will have values bounded by the upper and lower range limits.[8] By default, 5% of any normal healthy population will have values outside the normal range. Yet, it is unlikely that these subjects experience immune dysfunction of sufficient magnitude to cause disease. Often, it is difficult to determine whether aberrant subset determinations are biologi-cally relevant or a reflection of lymphocyte numbers in the small healthy population that falls out of the expected normal range.

In the interpretation of data from control and test groups, a probability value of 5% is often considered to be statistically significant. This is perfectly accept-able when only one variable is used in the experimental design. In studies to correlate exposure with alterations in lymphocyte subsets, the presence of numerous markers may be ascertained. When multiple markers are determined, there is an increased possibility that the frequency of one marker will be statistically different simply by chance alone. Therefore, there may be no biological significance ascribed to the observation. The Bonferroni inequality can be used to correct for this chance occurrence in multiple endpoint experiments.[35] If the probability value in classical tests equals or exceeds the p value divided by the number of markers tested, true statistical significance is insured. For example, when 20 markers are measured with a p value of 0.05%, the calculated significant value must equal or exceed a 0.0025 p value.

LYMPHOCYTE SUBSETS AND HEALTH-RELATED EFFECTS

Data from lymphocyte phenotyping studies cannot be extrapolated to prospectively determine disease or tumor susceptibility. The lack of a predictive capability is due to a complex matrix of theoretical, biological, and statistical factors. In theory, the immune system is charged with the defense and survival of the host against xenobiotics. The system has redundant effector loops and a functional reserve that must be exceeded before biological effects are noted.[36] Hence, minor perturbations in lymphocyte subsets have no biological relevance. To illustrate the functional reserve of the immune system, consider persons with primary immunodeficiencies or the acquired immunodeficiency syndrome (AIDS). Subjects with primary immunodeficiencies and susceptibility to select diseases have a virtual absence of specific immune functions. Also, data from studies on patients with secondary immunodeficiencies also suggest the existence of a large functional reserve. For example, the range of CD4 cells in normal peripheral blood is approximately 689–1683/mm^3. However, 50% of the newly diagnosed AIDS patients have CD4 cell numbers below 500/mm^3. An additional 12% have CD4 numbers below 200/mm^3. Yet, 50% of these subjects are asymptomatic and have no signs of infective processes or advanced HIV infections.[37]

The only antibodies having diagnostic capability are the CD3 and CD19 antibodies. These antibodies are licensed by the FDA for use in the diagnosis of leukemia and non-Hodgkins lymphoma—both malignant lymphoproliferative diseases that arise from single clones of cells. Because of the rapid division of malignant cells, increased numbers of lymphocytes will express pan T or B cell markers.[38] However, the biological matrix used in these flow cytometry studies is a bone marrow aspirate rather than a peripheral blood sample. In most cases, the test is used to support an already tentative clinical diagnosis.

UTILITY OF SPECIFIC ANTIBODIES AND LYMPHOCYTE SUBSETS AS BIOMARKERS

The determination of antigen-specific antibodies (IgG, IgM, or IgE) fulfills most of the criteria for an acceptable biomarker.[1] Antibodies are chemical specific, detectable in trace quantities, and relatable to prior exposure. The only disadvantage, however, is that blood for study must be drawn by invasive techniques.

If careful consideration is given to experimental design and proper control populations and exposure, data from flow cytometry analyses could be used as an index of exposure to specific chemicals. However, there are scant data in properly designed studies that support this concept. The use of lymphocyte phenotyping as a biomarker of exposure does, however, have merit. Both the Environmental Protection Agency (EPA) and the Center for Disease Control (CDC) are exploring the possibility of using lymphocyte phenotyping as biomarkers of exposure and both agencies have recommended the classical clinical study approach. More specifically, flow cytometry studies will be undertaken when there is clear clinical evidence that a xenobiotic is immunotoxic. The determination is made only after physical examinations are performed and exposure histories are clearly defined.

The utility of lymphocyte phenotyping has been demonstrated in subjects with medically diagnosed diseases. In patients with a well-documented medical history and objective signs or symptoms indicative of active disease, data from flow cytometry studies are useful in providing laboratory data that support a tentative clinical diagnosis. The data may, at the same time, provide information on immunological cells, active or impaired, in the disease process. For example, severely reduced numbers of CD4 cells are considered a hallmark of AIDS. Since CD4 cells are immunoregulatory cells involved in augmentation of the immune system, a decline in their number explains the unique pattern of infections observed in AIDS patients. Moreover, the number of CD4 cells is often used as an index to initiate therapeutic measures.[39] In other diseases such as multiple sclerosis or active lupus erythematosus (SLE), there is a decrease in the number of CD4,CD45R lymphocytes.[40,41] The reduction in the number of these cells, which amplify immunological suppression, may herald the activation of preexisting diseases with autoimmune components.

SUMMARY

Conceptually, the presence of xenobiotic induced IgG and IgM antibodies in serum can be used as an index of exposure while the presence of allergic antibody directed toward antigens may suggest health-related effects related to specific compounds. In practice, limited studies have been performed and only antibodies to some highly reactive small molecular weight chemicals have been demonstrated. Flow cytometry is an emerging technology with great scientific promise. Using this technology, lymphocyte subset phenotyping is easily achievable. However, the correlation between chemical exposure and altered lymphocyte populations is difficult to determine due to scientific problems associated with experimental design, and technical problems with the flow cytometer and the use of monoclonal antibodies. Finally, with the exception of altered numbers of CD3 and CD19 cells in the bone marrow, lymphocyte phenotyping has no predictive value or capability or correlation with health-related effects in normal, healthy subjects.

REFERENCES

1. Henderson, R. F., Bechtold, W. E., Bonds, J. A., and Sun, J. D. The use of biological markers in toxicology. *Crit. Rev. Toxicol.* 20, 65, 1989.
2. Mishell, B. B. and Shiigi, S. M. *Selected Methods in Cellular Immunology.* W. H. Freeman, San Francisco, 1980, p. 43.
3. Patterson, R., Roberts, M., Zeiss, C. R., and Pruzansky, J. J. Human antibodies against trimellityl proteins: Comparison of specificities of IgG, IgA and IgE classes. *Int. Arch. Allergy Appl. Immunol.* 66, 332, 1981.
4. Berstein, D. I., Patterson, R., and Zeiss, C. R. Clinical an immunological evaluation of trimellitic anhydride- and phthalic anhydride-exposed workers using a questionnaire with comparative analysis of enzyme linked immunosorbent and radioimmunoassays. *J. Allergy Clin. Immunol.* 69, 311, 1982.
5. Schlueter, D. P., Banazak, E. F., Fink, J., and Barboriak. J. Occupational asthma due to tetrachlorophthalic anhydride. *J. Occup. Med.* 20, 183, 1978.
6. Dykewicz, M. S., Patterson, R., Cugell, D. W., Harris, K. E., and Fang Wu, B. S. Serum IgE and IgG to Formaldehyde-human serum albumin: Lack of relation to gaseous formaldehyde exposure and symptoms. *J. Allergy Clin. Immunol.* 87, 48, 1991.
7. de Savingny, D. and Voller, A. Communication of ELISA data from laboratory to clinician. *J. Immunoassay* 1, 105, 1980. Mainland, D. Remarks on clinical "norms." *Clin. Chem.* 17, 267, 1971.
8. Reed, A. H, Henry, R. J., and Mason, W. B. Influence of the statistical method used on the resulting estimate of normal range. *Clin. Chem.* 17, 275, 1971.
9. McLaren, M., Drapper, C. C., Robert, J. M., Minter-Goedblood, E., and Lighthard, G. S. Studies on the enzyme-linked immunosorbent assay (ELISA) test for *Schistosoma mansoni* infections. *Ann. Trop. Med. Parasitol.* 72, 243, 1978.
10. Davis, D. D., Dulbecco, R., Eisen, H. N., Ginsberg, H. S., Wood, W. B. Jr., and McCarty, M. *Microbiology*, 2nd ed. Harper & Row, New York, 1973, p. 447.
11. Berg, T. L. O. and Johansson, S. G. O. Allergy diagnosis with the radioallergosorbent test. A comparison with the results of skin and provocation tests in an unselected group of children with asthma and hay fever. *J. Allergy Clin. Immunol.* 54, 207, 1974.
12. Lovett, E. J., Schnitzer, D. F., Karen, D. F., Flint, A., Hudson, J. L., and McClatchey, K. D. Applications of flow cytometry to diagnostic pathology. *Lab. Invest.* 50, 115, 1984.
13. Hensleigh, P. A., Bryan Water, B., and Henzenburg, L. Human T lymphocyte differentiation antigens: Effects of blood sample storage on the Leu antigens. *Cytometry* 3, 453, 1983.
14. Dunkberg, F. and Persidsky, M. D. A test of granulocyte membrane integrity and phagocytic function. *Cryobiology* 13, 430, 1976.
15. Patrick, C. W., Swartz, S. J., Harrison, K. A., and Keller, R. H. Collection and preparation of hematopoietic cells for membrane marker analyses. *Lab. Med.* 15, 659, 1984.

16. Miller, C. H. and Levy, N. B. Effect of storage conditions on lymphocyte phenotypes from healthy and diseased persons. *Clin. Lab. Anal.* 3, 296, 1989.

17. Knapp, M. S. and Pownall, R. Lymphocytes are rhythmic: Is it important? *Br. Med. J.* 298, 1328, 1984.

18. Levi, F., Cannon, C., Blum, P. J., Reinberg, A., and Mathe, G. Large amplitude circadian rhythm in helper:suppressor ratio of peripheral blood lymphocytes. *Lancet* ii, 462, 1983.

19. Abo, T., Kawate, T., Itoh, K., and Kumagi, K. Studies on the bioperiodicity of the immune response I. Circadian rhythms of human T, B, and NK cells in the peripheral blood. *J. Immunol.* 126, 1360, 1981.

20. Hallgren, H. H., Jackola, D. R., and O'Leary, J. J. Unusual pattern of surface marker expression on peripheral lymphocytes from aged humans suggestive of a population of less differentiated cells. *J. Immunol.* 131, 191, 1983.

21. Bradley, L. M., Bradley, J. S., Ching, D. L., and Shiigi, S. M. Predominance of T cells that express CD45R in the CD4+ helper/inducer lymphocyte subset of neonates. *Clin. Immunol. Immunopathol.* 51, 426, 1989.

22. Brahmi, Z., Thomas, J. E., Park, M., and Dowdswell, I. R. G. The effect of acute exercise on natural killer-cell activity of trained and sedentary human subjects. *J. Clin. Immunol.* 5, 321, 1985.

23. Chatenoud, L., Berrih, M. C., and Bene, T. The effect of immunomodulation on peripheral T cell subsets. *J. Clin. Immunol.* 2, 61S, 1982.

24. Westerman, J. and Pabst, R. Lymphocyte subsets: A diagnostic window on the lymphoid immune system. *Immunol. Today* 11, 406, 1990.

25. Dafoe, D. C., Stoolman, L. M., and Campbell, D. A. T cell subset patterns in cyclosporine treated renal transplant patients with primary cytomegalovirus disease. *Transplantation* 43, 452, 1987.

26. Roitt, I. *Essential Immunology*. Blackwell Scientific Publications, Boston, 1988, p. 163.

27. Biron, C. A., Byron, K. S., and Sullivan, J. L. Severe herpes virus infection in an adolescent without NK cells. *N. Engl. J. Med.* 320, 1731, 1989.

28. Routes, J. M. and Cook, J. L. Defective E1A gene product targeting of infected cells for elimination by natural killer cells. *J. Immunol.* 142, 4022, 1989.

29. Anderson, M., PaaBo, S., Nillson, T., and Peterson, P. A. Impaired intracellular transport of I MHC antigens as a possible means for adenovirus to evade immune surveillance. *Cell* 43, 215, 1985.

30. Jennings, S. R., Rice, P. L., Klosewiski, E. D., Anderson, R. W., Thompson, D. L., and Tevethia, S. S. Effect of herpes simplex virus types 1 and 2 on the surface expression of class I major histocompatibility complex antigens on infected cells. *J. Virol.* 56, 757, 1972.

31. Cook, J. L., Walker, T. A., Lewis, A. M., Ruley, H. E., Graham, F. L., and Pider, S. H. Expression of adenovirus E1A oncogene during transformation is sufficient to induce susceptibility to lysis by host inflammatory cells. *Proc. Natl. Acad. Sci. U.S.A.* 83, 6965, 1986.

32. Cook, J. L., May, D. L., Lewis, A. M., and Walker, T. A. Adenovirus E1A gene induction and susceptibility to lysis by natural killer cells and activated macrophages in rodent infected cells. *J. Virol.* 61, 3510, 1987.

33. Miller, L. G. and Ginns, L. C. Reversible defect in immunoregulating T cells in smoking. Analyses by monoclonal antibodies and flow cytometry. *Chest* 82, 526, 1982.

34. Phillips, B., Marshall, M. E., Brown, S., and Thompson, J. S. Effect of smoking on human natural killer cell activity. *Cancer* 56, 2789, 1985.

35. Miller, R. G. Jr. *Simultaneous Statistical Inferences,* 2nd ed. Springer-Verlag, New York, 1981, p. 62.

36. Dean, J. H., Thurmond, L. M., Lauer, L. D., and House, R. V. Comparative toxicology and correlative immunotoxicology in rodents. *Environmental Chemical Exposures and Immune System Integrity.* Bureger, E. J., Tardiff, R. G., and Bellanti, J. A., Eds. Princeton Scientific, Princeton, NJ, 1987, p. 85.

37. Hutchinson, C., Wilson, C., Reichart, C. A., Marsiglia, V. C., Zenilman, J. M., and Hook, E. W. CD4 lymphocyte concentrations in patients with newly identified HIV infections attending STD clinics. *JAMA* 266, 253, 1991.

38. Little, J. V., Foucar, K., Horvath, A., and Crago, S. Flow cytometric analyses of lymphoma and lymphoma like disorders. Seminar, *Diagnostic Pathol.* 6, 37, 1989.

39. Eyster, M. E., Ballard, J. O., Gail, M. H., Drummond, J. E., and Goedert, P. Predictive markers for the acquired immunodeficiency syndrome (AIDS) in hemophiliacs: Persistence of p24 and low T4 count. *Ann. Intern. Med.* 12, 863, 1989.

40. Morito, C., Hafler, H. L., and Weiner, J. Selective loss of suppressor induced T cell subsets in progressing multiple sclerosis: Analysis with anti-2H4 antibody. *N. Engl. J. Med.* 196, 307, 1987.

41. Raziuddin, S., Nur, M. A., and Alwabel, A. A. Selective loss of the CD4+ inducers of suppressor T cell subsets (2H4+) in active systemic lupus erythematosis. *J. Rheumatol.* 16, 315, 1989.

CHAPTER 9

Human Cytochromes P-450 and Their Roles in Environmentally-Based Disease

Frank J. Gonzalez

INTRODUCTION

Cytochromes P-450 (P-450s) are a large group of intrinsic membrane proteins ranging in size from 45 to 60 kDa and containing a single molecule of noncovalently bound heme.[1,2] Two main classes of P-450s exist, the xenobiotic-metabolizing P-450s and the steroidogenic P-450s. The latter are strictly involved in anabolic pathways leading synthesis of steroids such as testosterone, estrogen, cortisol, progesterone, and aldosterone and are produced in highly specialized tissues. This chapter will focus on xenobiotic-metabolizing P-450s. These P-450s have been referred to as mixed function monooxygenases because they add an atom of oxygen to numerous structurally diverse substrates. In the P-450 catalytic cycle, the enzyme finds substrate and the heme iron is reduced from a valency of +3 to +2 by an electron transferred from NADPH via another flavoprotein or iron sulfur protein electron carrier. O_2 then binds to the heme and is reduced by another electron. A series of reactions occur that result in splitting of O_2, production of H_2O, and oxidation of the substrate. Depending on the nature of the chemical substrate and rearrangements of intermediates at the active site, a number of reactions can occur as shown in Figure 9.1. With certain compounds the high energy intermediates can escape from the active site of the enzyme and react with cellular macromolecules such as protein, RNA, and DNA. These chemical substrates are usually promutagens, procarcinogens, or cellular toxins. P-450s can also inactivate or neutralize potentially harmful chemicals and drugs. Thus, these enzymes are necessary to convert foreign chemicals, that are frequently hydrophobic, to derivatives that can be easily eliminated from the body but, at the same time, can activate inert compounds to highly reactive metabolites.

Cytochrome P-450s have been classified based on primary amino acid sequence similarities.[3] Over 221 P-450 sequences have been determined from animals, plants, bacteria, and fungi, usually through cDNA sequences. By

0-87371-951-4/95/$0.00+$.50

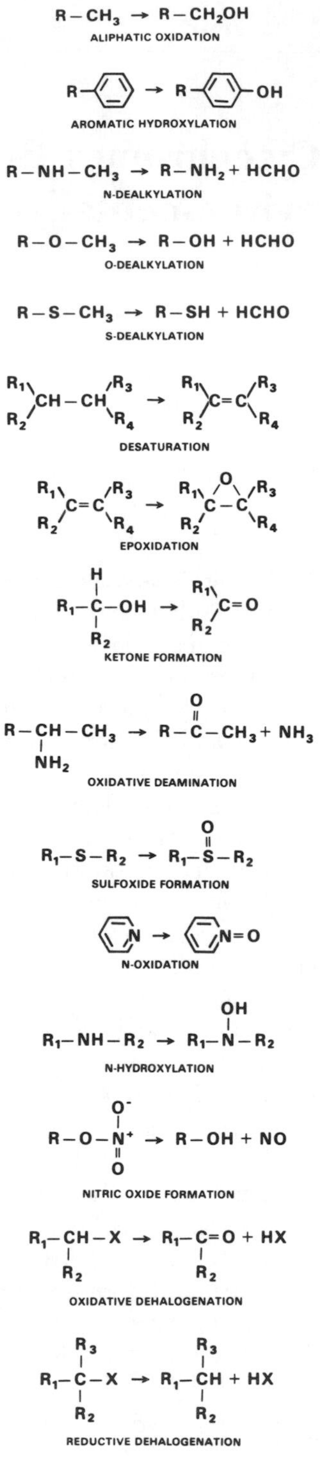

$$R-CH_3 \rightarrow R-CH_2OH$$
ALIPHATIC OXIDATION

$$R-\text{〇} \rightarrow R-\text{〇}-OH$$
AROMATIC HYDROXYLATION

$$R-NH-CH_3 \rightarrow R-NH_2 + HCHO$$
N-DEALKYLATION

$$R-O-CH_3 \rightarrow R-OH + HCHO$$
O-DEALKYLATION

$$R-S-CH_3 \rightarrow R-SH + HCHO$$
S-DEALKYLATION

$$\begin{array}{c} R_1 \\ \diagdown \\ R_2 \end{array} CH-CH \begin{array}{c} R_3 \\ \diagup \\ R_4 \end{array} \rightarrow \begin{array}{c} R_1 \\ \diagdown \\ R_2 \end{array} C=C \begin{array}{c} R_3 \\ \diagup \\ R_4 \end{array}$$
DESATURATION

$$\begin{array}{c} R_1 \\ \diagdown \\ R_2 \end{array} C=C \begin{array}{c} R_3 \\ \diagup \\ R_4 \end{array} \rightarrow \begin{array}{c} R_1 \\ \diagdown \\ R_2 \end{array} \overset{O}{\underset{}{C-C}} \begin{array}{c} R_3 \\ \diagup \\ R_4 \end{array}$$
EPOXIDATION

$$\begin{array}{c} H \\ | \\ R_1-C-OH \\ | \\ R_2 \end{array} \rightarrow \begin{array}{c} R_1 \\ \diagdown \\ R_2 \end{array} C=O$$
KETONE FORMATION

$$\begin{array}{c} R-CH-CH_3 \\ | \\ NH_2 \end{array} \rightarrow R-\overset{O}{\overset{\|}{C}}-CH_3 + NH_3$$
OXIDATIVE DEAMINATION

$$R_1-S-R_2 \rightarrow R_1-\overset{O}{\overset{\|}{S}}-R_2$$
SULFOXIDE FORMATION

$$\text{〇}N \rightarrow \text{〇}N=O$$
N-OXIDATION

$$R_1-NH-R_2 \rightarrow R_1-\overset{OH}{\overset{|}{N}}-R_2$$
N-HYDROXYLATION

$$R-O-\overset{O^-}{\underset{\overset{\|}{O}}{N^+}} \rightarrow R-OH + NO$$
NITRIC OXIDE FORMATION

$$\begin{array}{c} R_1-CH-X \\ | \\ R_2 \end{array} \rightarrow \begin{array}{c} R_1-C=O \\ | \\ R_2 \end{array} + HX$$
OXIDATIVE DEHALOGENATION

$$\begin{array}{c} R_3 \\ | \\ R_1-C-X \\ | \\ R_2 \end{array} \rightarrow \begin{array}{c} R_3 \\ | \\ R_1-CH \\ | \\ R_2 \end{array} + HX$$
REDUCTIVE DEHALOGENATION

Figure 9.1
Reactions catalyzed by
cytochromes P-450.

simultaneous alignments of sequences using computer programs, percentage amino acid sequence differences among P-450s are used to separate these enzymes into families and subfamilies. All P-450s within a single family display less than 40% sequence similarity with P-450s in any other family. All P-450s within a subfamily exhibit less than 59% sequence similarity with P-450s in any other subfamily within the same family. Only a few exceptions to these rules have been noted.[3]

Sequence similarity data also can be converted to accepted point mutations or PAMs as described by Dayhoff[4] and phylogenetic trees can be constructed.[3,5,6] These can be used to visualize the divergence patterns of P-450s and to speculate on the evolution of these proteins. For example, the oldest mammalian P-450s appear to be those involved in steroid metabolism while the most recent P-450s are those that metabolize xenobiotics.[5,6]

P-450s are named with the root CYP (*Cytochrome P-450*) followed by an Arabic number designating the family, a capital letter denoting the subfamily, and another Arabic number designating the individual P-450 number. For example, CYP1A1 is the number one P-450 in subfamily A of family 1. When two P-450s in the different species are believed to have clearly arisen from the same ancestor (termed orthologs) before divergence of the two species, they have the same P-450 number. Usually, the steroidogenic P-450s fall into this category. When orthologs are not identifiable P-450s in a subfamily, they are given separate numbers in different species.[3] Many P-450s in the CYP2 and CYP3 families have evolved in a species-specific manner so that their ancestral counterparts cannot be distinguished (see below).

The P-450 superfamily consists of 12 families in mammals.[3] Four families consisting of one or two P-450s encode the steroidogenic enzymes. Two families having single P-450s are involved in bile acid synthesis from cholesterol. A single subfamily designated CYP4A, containing several P-450s, encode fatty acid hydroxylases that are capable of oxidizing natural and synthetic fatty acids and drugs having long carbon chains. The main xenobiotic-metabolizing P-450s fall within the CYP1, CYP2, and CYP3 families. The CYP1 family contain one subfamily and two P-450s designated CYP1A1 and CYP1A2. The CYP2 family has seven subfamilies in mammals with one to several individual P-450s in each. The CYP3 family consists of two subfamilies, one of which contains a single P-450 expressed in olfactory epithelia, a cell that metabolizes foreign compounds.

A large degree of species differences occur in the catalytic activities of P-450s in the CYP2 and CYP3 families. Humans have P-450s with unique activities not found in rats or mice.[7] This is particularly important with respect to the carcinogen-activating P-450s since rodents or rodent-based systems are used to test chemicals under development for human use or exposure.

Human P-450s polymorphisms affecting drug metabolism have been found and analyzed at the molecular level. The most extensively studied is the debrisoquine/sparteine polymorphism in which about 7.5% of Caucasians are unable to metabolize a number of clinically used drugs.[8,9] A polymorphism in oxidation of the antiepileptic drug mephenytoin also exists, and is due to

Table 9.1   Substrates for Human P-450 Enzymes[a]

P-450 enzyme	Substrate
CYP1A1	Polycyclic aromatic hydrocarbons
CYP1A2	Imipramine, caffeine, phenacetin, verapamil, aflatoxins, arylamines, heterocyclic amines
CYP2A6	Aflatoxins, N-nitrosodiethylamine
CYP2B6	6-Aminochrysene
CYP2C8, CYP2C9	Benzophetamine, diazepam, diclofenac, hexobarbital, ibuprofen, imipramine, oxicam, antiinflammatory drugs, proguanil, propranolol, retinoic acid, S-warfarin, naproxen, tolbutamide
CYP2C19	S-Mephenytoin, omeprazole
CYP2D6	Antiarrhythmic agents, antihypertensives, β-blockers, monoamine oxidase inhibitors, morphine derivatives, antipsychotics, tricyclic antidepressants
CYP2E1	Chlorzoxazone
	Acrylonitrile, benzene, N-nitrosodimethylamine, vinyl halides, and other low-molecular-weight cancer suspect agents
CYP3A4, CYP3A5	Aldrin, benzphetamine, cyclosporin, erythromycin, lidocaine, lovastatin, midazolam, quinidine, 17α-ethynylestradiol, terfenadine, triazolam, various 1,4-dihydropyridines
	Aflatoxins, heterocyclic arylamines, nitroaromatic hydrocarbons
CYP4A11	Leukotriene receptor antagonists (long chain fatty acid hydroxylase)

[a] For each P-450 or subfamily both drug substrates and carcinogen/mutagen substrates are listed.

mutations in the *CYP2C19* gene.[10] A large degree of interindividual variation in expression of other P-450s, including those involved in carcinogen activation, exists in humans but it is unknown whether this variation has a genetic base or is due to the presence or absence of inducers found in the diet or introduced through drug administration. Indeed, expressions of rodent P-450s are markedly influenced by drugs.

METHODOLOGIES FOR STUDY OF HUMAN P-450s

Biochemical Approach

The human P-450s responsible for metabolism of foreign compounds are listed in Table 9.1 along with their activities toward drugs, promutagens, and procarcinogens. These data are largely summarized by Guengerich and Shimada,[11] and Gonzalez.[12] Specific references and more detailed discussions can be found in these articles.

Human P-450s have been studied by direct purification and analysis of catalytic activities after reconstitution in lipid vesicles with NADPH-P-450 oxidoreductase. In many cases, however, the P-450s are not fully active due to the artificial nature of the reconstituted system or because the P-450s are partially denatured. Antibodies prepared against individual P-450 forms can be used to assess activities in total human liver endoplasmic reticulum membranes, commonly called microsomes because of their vesicular nature. In one type of experiment the antibodies are added to the microsomes prior to carrying out an enzyme assay. If the P-450 to which the antibody binds is involved in a catalysis of the substrate, microsomal activity will be decreased (Fig. 9.2A, curve 1). If

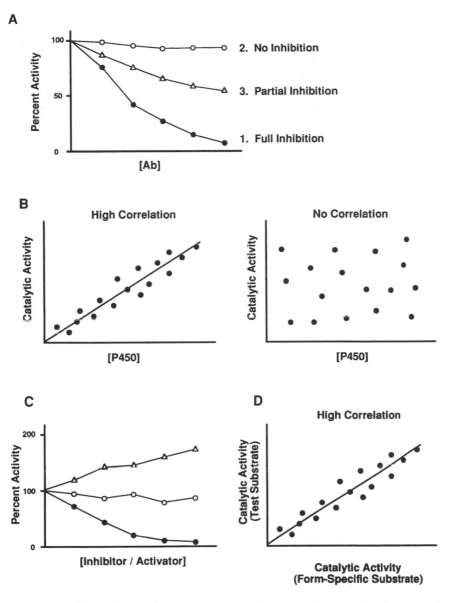

Figure 9.2 (A–D) Analysis of catalytic activities of human P-450s using biochemical and immunocorrelative approaches.

the P-450 is not involved, no difference will be detected when compared to a control assay preparation that had been preincubated with nonimmune antibody (Fig. 9.2A, curve 2). If two different P-450s, one of which does not react with the antibody, are involved in the reaction, partial inhibition is observed (Fig. 9.2A, curve 3).

Immunocorrelation studies are also carried out to determine whether a certain P-450 form catalyzes a reaction. In this case, a collection of human liver microsome samples is used. Human livers have usually been obtained from kidney donor cadavers, accident victims, and in some cases surgery. Levels of specific P-450 forms are quantified by Western immunoblot analysis. Usually, a large degree of difference exists in P-450 levels between different livers, so a wide variation in P-450 content is found. Activity measurements, preferably V_{max} values, are then made on the different microsomes, and the rate of product formation and level of P-450 for each liver are plotted (Fig. 9.2B). Linear regression analysis with a high correlation coefficient indicates that the P-450 to which the antibody was used for immunoquantitation is responsible for carrying out the reaction. If the points are scattered, the P-450 in question is not involved or more than one P-450 is involved. One drawback to these studies described in Figure 9.2A and B is that the antibodies must be highly specific. In many cases, P-450s share high sequence similarities, therefore, making specific antibodies is difficult. It can also be difficult experimentally to demonstrate specificity.

A third method that has been used to determine P-450 catalytic specificity is form-specific chemical inhibitors or activators.[11] For example, 7,8-benzoflavone inhibits activities of CYP1A1 and CYP1A2 P-450s and stimulates those of the CYP3A P-450s. A typical experiment showing activation, no effect, and inhibition, is shown in Figure 9.2C.

A fourth method of determining whether a specific reaction is catalyzed by a single P-450 form is by activity correlations.[11] Once a specific substrate is found for a particular P-450 form, activities can be measured for several human liver microsome samples. Other substrates can be tested using the same microsome specimens and their rates of metabolism are plotted vs. rates of metabolism of the form-specific substrate. The data are analyzed by linear regression (Fig. 9.2D).

For examination of activation of promutagens and procarcinogens a bacterial DNA damage assay has been used.[11] This straightforward assay allows the examination of most compounds except the nitrosamines. Thus, with a standard set of human liver microsome samples with known activities toward form-specific substrates, numerous unknown drugs promutagens and procarcinogens can be analyzed to determine which P-450 forms carry out a specific reaction.

Molecular Biology Approach

Another method that has been used to study catalytic activities of human P-450s is cDNA expression.[7] Complementary DNAs (cDNAs) corresponding to individual P-450 forms can readily be isolated from cDNA libraries prepared from RNA isolated from human tissue specimens (Fig. 9.3A). Messenger RNA is prepared from liver or lung tissue samples and converted to double-stranded cDNAs by use of reverse transcriptase. These cDNAs should represent all mRNAs in a given tissue. The cDNAs are then inserted into prokaryote expression vectors in the form of plasmids or viruses. These are introduced into bacteria to generate a "cDNA library," which is screened using antibodies against rat or human P-450s or by rat cDNA probes. Over 100,000 different cDNAs can be easily screened. Specific P-450 cDNAs are isolated and sequenced to obtain the complete amino acid reading frame of the enzyme.

The cDNA can be used to produce catalytically active P-450 by use of cDNA expression (Fig. 9.3B). A number of different systems have been used including yeast, COS cell-SV40, vaccinia virus, and human B-lymphoblastoid cells.[2,7,11] The cDNA is introduced into a vector that contains a strong RNA polymerase II promotor. The vector is then introduced into cultured cells. Cells containing the vector are selected by treatment with a drug that kills only cells that do not contain the plasmid. The expression plasmid contains a gene encoding a protein that degrades the drug. In the case of vaccinia virus expression systems, cells are infected with lytic virus containing the cDNA and are killed within a few days. During the lytic cycle, large amounts of active P-450 are produced. Thus, cDNA expression allows one to analyze individual P-450 catalytic activities without purifying the enzyme.

The immunochemical and cDNA expression approaches can yield complementary data on a human P-450 catalytic activities. Individually expressed P-450s can be used to determine the range of substrates for a particular P-450, including form-specific substrates and substrates that are metabolized by two or more P-450 forms. Immunochemical analysis of liver microsomes can be used to determine the role of individual P-450 forms in metabolism of a particular substrate in an environment containing all liver P-450s.

HUMAN XENOBIOTIC-METABOLIZING P-450s

CYP1A1 and CYP1A2

CYP1A1 was identified as the enzyme responsible for metabolic activation of polycyclic aromatic hydrocarbon carcinogens,[13] found in polluted environments and in cigarette smoke. This activity is well conserved in mammals. CYP1A1 is not appreciably expressed in the absence of inducing agents. Most substrates for the enzyme induce its synthesis through a receptor–ligand-mediated transcriptional activation.[14] Expression of CYP1A1 has been detected in lung[15] and placenta[16] of smokers but has not been seen in most liver specimens including those of smokers.[17,18] The lack of CYP1A1 expression in livers of smokers may be due to the fact that these livers were obtained from surgical specimens or through kidney donor programs long after the individual smoked a cigarette. Indeed CYP1A1 expression in lungs decays as a function of time after smoking cessation.[15] Although expression of CYP1A1 has been linked to cancer in rodents,[13] no definitive evidence exists suggesting an association in humans. A correlation has been found between extent CYP1A1 activity in lymphocytes and lung cancer.[19] But this has yet to be confirmed as a genetic association. A restriction fragment length polymorphism in the *CYP1A1* gene has been associated with increased risk of lung cancer in Japanese smokers.[20,21]

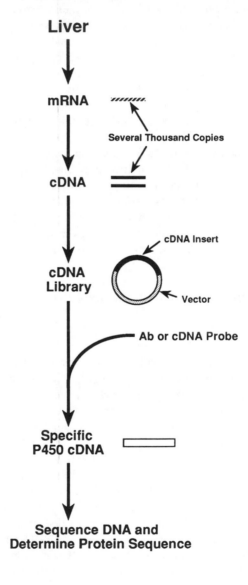

Figure 9.3 (A, B) Analysis of P-450 catalytic activities through cloning, sequencing, and expression of cDNAs.

B.

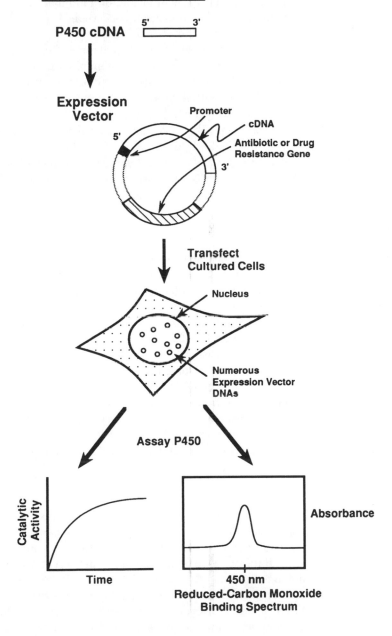

Figure 9.3 *(Continued.)*

The same polymorphism was found in a Norwegian population but it was not associated with increased incidence of lung cancer.[22]

The CYP1A2 gene is constitutively expressed in human liver [17,18] but not extrahepatic tissues. In rodents, CYP1A2 is inducible by a number of different agents including dioxin and polycyclic aromatic hydrocarbons.[2] It is unknown whether this enzyme is also inducible in humans.

CYP1A2 metabolizes a large number of compounds including drugs and carcinogens. Caffeine is N-demethylated[23] and acetaminophen is activated to quinone toxic metabolites by CYP1A2.[24] The drug phenacetin is also metabolized by this enzyme.[25] Most importantly, CYP1A2 metabolically activates arylamine carcinogens such as 2-acetylaminofluorene, 2-aminofluorene, 2-naphthylamine, 2-aminoanthracene, 4-aminobiphenyl, and all known heterocyclic amine promutagens derived from cooked beef.[11] These include 2-amino-3-methylimidazo[4,5-f]quinoline (IQ)[26] and 2-amino-1-methyl-6-phenylimidazo[4,5-b]pyridine (PhIP).[27] A more complete list of substrates and inhibitors of CYP1A2 can be found in a recent review.[11]

CYP2A6

CYP2A6 was first identified in human liver by cDNA cloning experiments.[28,29] The enzyme was later purified from a liver specimen.[30] A P-450 within the same subfamily and having similar enzymatic characteristics designated CYP2A3 is expressed in rat lung.[31] However, it is unknown whether CYP2A6 is found in human pulmonary tissue or other human extrahepatic tissues or whether it is inducible in human. The rat CYP2A P-450s can be induced by 3-methylcholanthrene.[32] There is a large degree of variation in levels of expression of hepatic CYP2A6 in human liver specimens.[28,29] Some livers have high activities whereas others possess low or undetectable activities.

CYP2A6 is probably the only human P-450 that carries out coumarin 7-hydroxylation.[28] Specificity of this reaction on a relatively innocuous compound may be useful for *in vivo* monitoring of CYP2A6 levels. The enzyme is also capable of activating the procarcinogen aflatoxin B_1 [32] and N-nitrosodiethylamine.[33]

CYP2B6

CYP2B6 was never isolated from human tissue but its cDNA was cloned and the active enzyme expressed.[34] This enzyme is expressed in human liver and similar to CYP2A6, a large degree of interindividual differences in level of expression were found. Interestingly livers containing high CYP2A6 mRNA levels were also found to contain high CYP2B6 mRNA levels.[28,29,34] These data suggest that certain patients may have been exposed to inducers that regulate the CYP2A6 and CYP2B6 genes. However, no evidence for inducers was found in the medical reports accompanying the livers, most of which were obtained from kidney donors. The expression of CYP2B6 in extrahepatic tissues has not been thoroughly investigated. It would not be surprising to find expression in other tissues since rodent *CYP2B* gene expressions are found in lung,[35,36] intestine,[37] and testis.[36]

CYP2B6 was found to metabolically activate aflatoxin B_1,[32] the chemotherapeutic agent cyclophosphamide (C.L. Crespi and F.J. Gonzalez, unpublished), and nicotine.[35] No specific substrates for this enzyme have been found that are not activated to toxic metabolites. CYP2B6 can deethylate 7-ethoxycoumarin, however, this compound is a substrate for several other human P-450 forms. Thus, currently no candidate substrate has been found that is suitable for *in vivo* monitoring of CYP2B6.

CYP2C Subfamily

Six human P-450s have been identified in this subfamily and designated CYP2C8, CYP2C9, CYP2C10, CYP2C17, CYP2C18, and CYP2C19.[3] Only CYP2C8 and CYP2C9 have been purified[39-43] while the remaining were found by cDNA cloning. CYP2C19 is responsible for the *S*-mephenytoin polymorphism in which 2–5% of Caucasians and about 25% of Japanese are deficit in metabolism of this antiepileptic drug.[10] None of the other CYP2C P-450 forms have demonstrated any stereoselective metabolism of the *S*-isomer of the drug, although some hydroxylate the *R*-isomer.[44,45] A CYP2C9 variant isolated from a Japanese liver specimen, which was found to catalyze low rates of *S*-mephenytoin 4'-hydroxylation.[46]

CYP2C8 and CYP2C9 both catalyze oxidation with the hypoglycemic drug tolbutamide.[44,46] CYP2C9 is also the principal enzyme responsible for metabolism of the anticoagulant drug warfarin by hydroxylation at the 7 carbon position on the molecule.[47] The 7-hydroxywarfarin metabolite is the most abundant found in urine of patients receiving the drug. Other human P-450s also hydroxylate warfarin at different carbon positions but these metabolites are not found in high levels in human plasma and urine after administration with the drug. To date no promutagens or procarcinogens have been found that are metabolized by CYP2C P-450s.

CYP2D6

CYP2D6 is one of the best-studied P-450s in the CYP2 family. This enzyme is responsible for the debrisoquine/sparteine drug oxidation polymorphism. Over 25 drugs are known whose metabolisms cosegregate with the debrisoquine/sparteine genetic polymorphism.[8] These include the analgesic codeine, which is activated by CYP2D6 through an *O*-demethylation reaction to morphine, the cardiovascular drugs encainide and flecainide, the tricyclic antidepressants amitriptyline, nortriptyline and desmethylimipramine, and the ß-adrenergic blocking drugs metaprolol and propranalol.

The debrisoquine/sparteine genetic polymorphism is due to the presence of mutant *CYP2D6* genes. Five mutant alleles have been found and characterized that account for over 98% of all mutant genes. The major allele representing about 75% of all mutants contains a G → A base change at the junction of intron 4 and exon 5 of the *CYP2D6* gene that alters the consensus splice site of AG to an AA. The second most common defective allele is due to a complete deletion of the *CYP2D6* gene. On the basis of the molecular characterization of the *CYP2D6* mutant genes a polymerase chain reaction procedure has been developed that can be used to genotype patients eliminating the need for the classical phenotyping test.[48,49]

CYP2E1

CYP2E1 is the only P-450 capable of metabolizing ethanol. It carries out oxidation of acetone and acetol, reactions that are associated with a gluconeogenic pathway of intermediary metabolism.[50] This P-450 and its associated activities are well conserved in mammals indicating that it may be required for normal physiologically-based metabolic pathways as opposed to an exclusive role in xenobiotic metabolism.

CYP2E1 metabolizes the drugs acetaminophen, chlorzoxazone, and halothane and the solvents benzene, carbon tetrachloride, dichloromethane, pentane, ethylcarbamate, acrylonitrile, vinyl chloride, styrene, chloroform, and pyridine. These reactions frequently yield toxic metabolites capable of damaging cellular macromolecules and producing cell toxicity. CYP2E1 also metabolically activates the potent procarcinogen *N*-dimethylnitrosamine.

CYP2E1 is expressed at highest level in the liver and is induced in this tissue and other extrahepatic tissues, including lymphocytes,[51] by many of its own substrates by a post-transcriptional mechanism.[52] Induction is due to stabilization of the enzyme against degradation. It is unclear, however, whether this is due to allosteric binding of the substrate or to continued occupancy of the active site.

It has been proposed that the muscle relaxant drug chlorzoxazone be used as an *in vivo* monitor of CYP2E1 expression.[53] This compound can be used in a way that debrisoquine and sparteine were used to determine CYP2D6 levels in humans by measuring levels of 6-hydroxychlorzoxazone in urine to determine the extent of expression of CYP2E1.

CYP2F1

CYP2F1 was identified through cDNA cloning experiments,[54] and has never been directly purified. This P-450 is expressed in lung and possibly other extrahepatic tissues but is not expressed to any significant degree in liver. cDNA-directed expression was used to analyze the catalytic activity of CYP2F1. The enzyme metabolically activates the pneumotoxin naphthylamine (K. Korzekwa and F. Gonzalez, unpublished results).

CYP3A Subfamily

At least four P-450s are expressed in the CYP3A subfamily.[3] CYP3A4 and CYP3A3 exhibit only 13 amino acid residue differences and are probably expressed in most adult human livers although their extent of expressions differ significantly.[55] No significant differences have been detected in the substrate specificities between these two enzymes. Another P-450, designated CYP3A5, was found that displays about 84% amino acid sequence similarity to CYP3A3 and CYP3A4. Interestingly, CYP3A5 is expressed in only about 20–30% of adult livers. It is unknown, however, whether this variability of expression is due to mutant CYP3A5 genes. Only quantitative differences have been detected in catalytic activities between CYP3A5 and CYP3A3/CYP3A4 except for the substrate cyclosporine; CYP3A3 and CYP3A4 produce three distinct metabolites of this cyclic peptide whereas CYP3A5 yields only one of these metabolites.[56]

CYP3A3, CYP3A4, and CYP3A5 metabolize numerous drugs including cyclosporine, erythromycin, ethynylestradiol, lidocaine, midazolam, nifedipine, warfarin, and quinidine. They also hydroxylase the natural steroids testosterone and cortisol. CYP3A3/CYP3A4 are also capable of activating the human hepatocarcinogens aflatoxin B_1.[11] Immunochemical evidence suggests involvement in activation of aflatoxin G_1, sterigmatocystin, senecionine, procarcinogenic metabolites of benzo[a]pyrene and dibenz[a]anthracine, 6-aminochrysene, and 1-nitropyrene.[11]

A P-450, designated CYP3A7, has been identified in fetal liver.[56–58] This P-450, which has not been detected in adult livers, is probably the major P-450 in fetal liver. This contrasts with rodents which do not significantly express P-450s in fetal liver in the absence of inducers. CYP3A7 catalyzes dehydroepiandosterone 3-sulfate 16α-hydroxylation in the metabolic pathway leading to estriol formation during pregnancy.[59] This may be a physiologically important role of CYP3A7. This enzyme is also capable of hydroxylating testosterone at the 6ß position and probably other xenobiotic substrates. Thus CYP3A7 may have a dual role of steroid synthesis and foreign compound metabolism in the fetus.

Due to the similarity in substrate specificities between the adult CYP3A P-450s, it is not possible to monitor their individual levels by use of *in vivo* probes. However, nifedipine, erythromycin, dapsone, lidocaine, and cortisol have been proposed as noninvasive compounds that may be useful to monitor the CYP3A P-450s as a class.[11]

CYP4B1

CYP4B1 was identified by cDNA cloning from a human lung cDNA library.[63] This P-450 is not appreciably expressed in human liver but is probably expressed in other extrahepatic tissues. An ortholog of CYP4B1 has been extensively studied in rabbits that activates the carcinogen 2-aminofluorene and 4-ipomeanol.[61–63] Human CYP4B1, however, is unable to activate either of these chemicals, indicating a marked species difference in catalytic activity. Human CYP4B1 catalyzes specific hydroxylations of testosterone and estradiol. These reactions are similar to those carried out by the CYP3A P-450s, except that CYP4B1 cannot activate aflatoxin B_1.[32] cDNA-directed expression is currently being used to determine other substrates for this enzyme.

ROLE OF P-450 IN ENVIRONMENTALLY BASED DISEASE

Metabolism and Disposition of Xenobiotics

Almost all foreign chemicals when ingested or absorbed by the body are metabolized by the so called "drug-metabolizing" enzymes. These include the phase I or functionalizing enzymes such as P-450s and the flavin-containing monooxygenases and the phase II or conjugating enzymes such as UDP-glucuronosyltransferases, sulfotransferases, glutathione transferases, methyltransferases, and N-acetyltransferases. A number of other xenobiotic-metabolizing enzymes exist that do not strictly fall into either class; these include carboxylesterases, aldehyde dehydrogenases, epoxide hydrolases, dihydrodiol dehydrogenases, NAD(P)H-quinone oxidoreductase, and NADPH-P-450 oxidoreductase, the latter of which serves to donate electrons to P-450 and to directly reduce aldehydes and quinones.

In a typical set of reactions, a hydrophobic molecule will enter the body, get transferred from the intestine or lungs to the liver, and there it will diffuse into the lipid bilayer of the hepatocyte intracellular membranes or endoplasmic reticulum. Here the compound is metabolized by the P-450s into a number of potential products, the most common and innocuous being a hydroxy derivative. This "functionalized" metabolite can then serve as a substrate for one or more of the transferases, which add a hydrophilic constituent chemical. This converts the chemical to a water-soluble form that can then be easily eliminated from the body through urine or bile.

Unfortunately, many chemicals are metabolically activated by P-450s to toxic or carcinogenic metabolites, usually electrophiles, that bind to cellular macromolecules. In some cases, however, these high-energy metabolites can be scavenged or conjugated by the transferases. Certain epoxides can also be substrates for epoxide hydratases thus neutralizing their reactivities. At high enough concentrations many chemicals can kill the cells resulting in toxicity. For example, acetaminophen, a commonly used over-the-counter analgesic, is converted to cell-killing quinones by CYP1A2 or CYP2E1.[24] Most chemical carcinogens enter the body as inert compounds and P-450s covert them to reactive species that bind DNA. It is generally believed that certain "DNA adducts" are inefficiently repaired and lead to gene mutations, the most critical being oncogenes and/or tumor suppressor genes.[67]

P-450 and Cancer

It is well established that there exists a strong genetic component to chemical carcinogenesis in many tissues, including breast and colon.[64] Susceptibility to lung cancer has also been inherited and is not solely due to environmental factors such as smoking.[65]

Since P-450s are responsible for procarcinogen activation it remains a possibility that interindividual differences in their expressions play a role in cancer susceptibility or resistance. Indeed evidence in rodents clearly suggests a role for P-450s as noted in the cancer incidences of mice strains that differ in their inducibilities for CYP1A1, the polycyclic aromatic hydrocarbon metabolizing P-450.[13] There is no firm link between P-450 expression and human cancer, although a number of provocative suggestive associations have been published.

CYP1A1

CYP1A1 is highly induced in lungs of smokers[15] and as such could play a role in lung cancer risk. Early studies demonstrated two populations of high and low inducibility phenotypes as assessed by analysis of lymphocytes.[66] Others could not reproduce these results but demonstrated a large variability in activity dependent on a number of factors.[67] A correlation was later found between high constitutive expression of CYP1A1 activity and smoking associated with lung cancer.[68] These studies were carried out under well-controlled conditions and the activities were normalized to NADPH-P-450 oxidoreductase levels. However, another complete study on a separate group of patients and controls has not been done.

A restriction fragment length polymorphism (RFLP) was found in the human *CYP1A1* gene that results from a mutation in the restriction enzyme *Msp*I recognition sequence located downstream of the gene.[20] This base change apparently does not affect the function of the gene. The allele frequencies in a population of 375 Japanese individuals was 0.67 and 0.33 for the common, designated m1 (no *Msp*I site) and rare, designated m2, alleles.[21] Analysis of lung cancer patients and case controls revealed that individuals of the m2/m2 genotype were at higher risk for lung cancer, in particular risk toward development of the Kreyberg type I cancer. In addition m2/m2 individuals contracted lung cancer with a significantly lower cumulative doses of cigarette than the m1/m1 and m1/m2 genotypes. It was also found that the m2 homozygous genotype confired a much higher relative risk for lung cancer when it occurred in an individual that was also deficient in a glutathione S-transferase designated GSTM1.[69]

As noted, the base change causing these RFLP is not associated with any structural or apparent regulatory region of the gene. It is in linkage disequilibrium with an amino acid change in exon 7 of CYP1A1 in which an Ile in the major allele is replaced by a Val.[69] The Msp I polymorphism may also be in linkage disequilibrium with a base change that is crucial for gene transcription or P-450 activity. This has yet to be investigated.

CYP2D6

A number of case control studies have been accomplished analyzing the relationship between the debrisoquine/sparteine polymorphism and lung cancer.[70] Lung cancer patients were almost always of the extensive metabolizing phenotype.[71] Further analysis of these data revealed that all nonadenocarcinoma patients were of the extensive metabolizing phenotype.[72] Consideration of other occupational exposures yielded increased risk of up to 18-fold in combination with debrisoquine/sparteine metabolic phenotypes. This association was confirmed by a study on a separate population of individuals,[73] but was not found in two separate studies.[44,74] Possible reasons for these differences have been discussed.[75-78]

The biochemical basis for the association of debrisoquine/sparteine metabolic phenotype with lung cancer is not known. CYP2D6 activates these tobacco-specific nitrosamine 4-(methylnitrosamino)-1-(3-pyridyl)-1-butanone (NNK) to toxic and mutagenic metabolites, however, other human P-450s such as CYP1A2 and CYP2A6 and to a lesser extent CYP2E1 are also capable of activating this procarcinogen.[78] It is not known whether CYP2D6 is expressed in human lung.

Future Prospects

The role of human P-450s in environmentally based disease should now be amenable to study given the solid biochemical foundations that have been laid during the past few years in determining the number and complexity of human P-450 forms and their substrate specificities. Noninvasive chemicals or "probes" that specifically measure *in vivo* P-450 catalytic activities are being developed to determine levels of expression of individual P-450 forms. It might be possible to administer a single tablet containing several compounds to people to simultaneously measure levels of different P-450 forms. In some cases polymerase chain reaction-based procedures can be used to determine presence of mutant P-450 genes from leukocyte DNAs. By use of these assays, molecular epidemiological studies, both cohort, case control, and prospective, can be used to determine whether interindividual variations in levels of any P-450 forms are associated with disease such as cancer or neurological disorders.

REFERENCES

1. Gonzalez, F. J. Cytochromes P-450. *Encyclopedia of Human Biology*. Dulbecco, R., Ed. Academic Press, San Diego, 1991, Vol. 2, p. 737.
2. Gonzalez, F. J. The molecular biology of cytochrome P-450s. *Pharmacol. Rev.* 40, 243, 1988.
3. Nelson, D. R., Kamataki, T., Waxman, D. J., Guengerich, F. P., Estabrook, R. W., Feyereisen, R., Gonzalez, F. J., Coon, M. J., Gunsalus, I. C., Gotoh, O., Okuda, K., and Nebert, D. W. The P 450 superfamily: Update on new sequences, gene mapping, accesion numbers, early trivial names of enzymes, and nomenclature. *DNA Cell Biol.* 12, 1, 1991.
4. Dayhoff, M. O. *Atlas of Protein Sequence and Structure*, Vol. 5, Suppl. 3. National Biomedical Research Foundation, Silver Spring, MD.
5. Nelson, D. R. and Strobel, H. W. Evolution of cytochrome P-450 proteins. *Mol. Biol. Evol.* 4, 572, 1987.
6. Nebert, D. W., Nelson, D. R., and Feyereison, R. Evolution of the cytochrome P-450 genes. *Xenobiotica* 19, 1149, 1989.

7. Gonzalez, F. J., Crespi, C. L., and Gelboin, H. V. cDNA-expressed human cytochrome P 450: A new age of molecular toxicology and human risk assessment. *Mutat. Res.* 247, 113, 1991.
8. Eichelbaum, M. and Gross, A. S. The genetic polymorphism of debrisoquine/ sparteine metabolism—clinical aspects. *Pharmac. Ther.* 46, 377, 1990.
9. Gonzalez, F. J. and Meyer, U. A. Molecular genetics of the debrisoquine polymorphism. *Clin. Pharm. Ther.* 50, 233, 1991.
10. Goldstein, J. A. and de Morais, S. M. F. Biochemistry and molecular biology of the human CYP2C subfamily. *Pharmacogenetics* 4, 285, 1994.
11. Guengerich, F. P. and Shimada, T. Oxidation of toxic and carcinogenic chemicals by human cytochrome P-450 enzymes. *Chem. Res. Toxicol.* 4, 391, 1991.
12. Gonzalez, F. J. P-450s in humans. *Cytochrome P-450.* Schenkman, J. B., and Greim, H., Eds. *Handbook of Experimental Pharmacology.* Springer-Verlag, Heidelberg, Germany, 1993, p. 239.
13. Nebert, D. W. The Ah locus: Genetic differences in toxicity, cancer, mutation and birth defects. *Current Rev. Toxicol.* 20, 153, 1989.
14. Swanson, H. I. and Bradfield, C. A. The Atl-receptor: genetics, structure, and function. *Pharmacogenetics* 3, 213, 1993.
15. McLemore. T., Adelberg, S., Lim, M. C., McMahon, N. A., Yu, S. J., Hubbard, W. C., Czerwinski, M., Coudert, B. P., Moscow, J. A., Stinson, S., Storeng, R., Lubert, R. A., Eggleston, J. C., Boyd, M. R., and Hines, R. W. Cytochrome P-4501A1 gene expression in lung cancer patients: Evidence for cigarette smoke-induced expression in normal lung and altered gene regulation in primary pulmonary carcinomas. *J. Natl. Cancer Inst.* 82, 1333, 1990.
16. Song, B. J., Friedman, F. K., Park, S. S., Tsokos, G. C., and Gelboin, H. V. Monoclonal antibody-directed radioimmunoassay detects cytochrome P-450 in human placenta and lymphocytes. *Science* 228, 490, 1985.
17. Wrighton, S. A., Campanile, C., Thomas, P. E., Maines, S. L., Watkins, P. B., Parker, G., Mendez-Picon, G., Haniu, M., Shively, J. E., Levin, W., and Guzelian, P. S. Identification of a human liver cytochrome P-450 homologous to the major isosafrole-inducible cytochrome P-450 in the rat. *Mol. Pharmacol.* 29, 405, 1986.
18. McManus, M. E., Stupans, I., Ioanni, B., Burgess, W., Rabson, R. A., and Birkett, D. J. Identification and quantitation in human liver of cytochrome P-450 analogous to rabbit cytochromes P-450 forms 4 and 6. *Xenobiotica* 18, 207, 1988.
19. Kouri, R. E., McKinney, C. E., Slomiany, D. J., Snodgrass, D. R., Wray, N. P., and McLemore, T. L. Positive correlation between high aryl hydrocarbon hydroxylase activity and primary lung cancer as analyzed in cryopreserved lymphocytes. *Cancer Res.* 42, 5030, 1982.
20. Kawajiri, K., Nakachi, K., Imai, K., Yoshii, A., Shinoda, N., and Watanabe, J. Identification of genetically high risk individuals to lung cancer by DNA polymorphisms of the cytochrome P-4501A1 gene. *FEBS Lett.* 263, 131, 1990.
21. Kawajiri, K., Nakachi, K., Imai, K., Hayashi, S., and Watanabe, J. The individual difference of susceptibility to lung cancer in relation to polymorphisms of the P-4501A1 gene and cigarette dose. *Xenobiotics and Cancer.* Ernster, L., Gelboin, H. V., and Sugimura, T., Eds. Japan Society Press, Tokyo, 1991, pp. 55–61.
22. Tefre, T., Ryberg, D., Haugen, A., Nebert, D. W., Skaug, V., Brogger, A., and Borresen, A. L. Human CYP1A1 (cytochrome $P_1 450$) gene: Lack of association between the Msp I restriction fragment length polymorphism and incidence of lung cancer in a Norwegian population. *Pharmacogenetics* 1, 20, 1991.

23. Butler, M. A., Iwasaki, M., Guengerich, F. P., and Kadlubar, F. F. Human cytochrome P-450$_{PA}$ (P-450IA2), the phenacetin O-deethylase, is primarily responsible for the hepatic 3-demethylation of caffeine and N-oxidation of carcinogenic arylamines. *Proc. Natl. Acad. Sci. U.S.A.* 86, 7696, 1989.

24. Raucy, J. L., Lasker, J. M., Lieber, C. S., and Black, M. Acetaminophen activation by human liver cytochrome P-450IIE1 and P-450IA2. *Arch. Biochem. Biophys.* 271, 270, 1989.

25. Distlerath, L. M., Reilly, P. E. B., Martin, M. V., Davis, G. G., Wilkinson, G. R., and Guengerich, F. P. Purification and characterization of the human liver cytochromes P-450 involved in debrisoquine 4-hydroxylation and phenacetin O-deethylation, two prototypes for genetic polymorphism in oxidative drug metabolism. *J. Biol. Chem.* 260, 9057, 1985.

26. Aoyama, T., Gelboin, H. V., and Gonzalez, F. J. Mutagenic activation of 2-amino-3-methylimidazo [4,5-*f*] quinoline by complementary DNA-expressed human liver P-450. *Cancer Res.* 50, 2060, 1990.

27. McManus, M. E., Burgess, W. M., Veronese, M. E., Huggett, A., Quattrochi, L. C., and Tukey, R. H. Metabolism of 2-acetylaminofluorene and benzo (a) pyrene and activation of food-derived heterocyclic amine mutagens by human cytochromes P-450. *Cancer Res.* 50, 3367, 1990.

28. Yamano, S., Tatsuno, J., and Gonzales, F. J. The CYP2A3 gene product catalyzes coumarin 7-hydroxylation in human liver microsomes. *Biochemistry* 29, 1322, 1990.

29. Miles, J. S., McLaren, A. W., Forrester, L. M., Glancey, M. J., Lang, M. A., and Wolf, C. R. Identification of the human liver cytochrome P-450 responsible for coumarin 7-hydroxylase activity. *Biochem. J.* 267, 365, 1990.

30. Yun, C. H., Shimada, T., and Guengerich, F. P. Purification and characterization of human liver microsomal P-4502A6. *Mol. Pharmacol.* 40, 679, 1992.

31. Kimura, S., Kozak, C. A., and Gonzalez, F. J. Identification of a novel P-450 expressed in rat lung: cDNA cloning and sequence, chromosome mapping, and induction by 3-methylcholanthrene. *Biochemistry* 28, 3798, 1989.

32. Aoyama, T., Yamano, S., Guzelian, P. S., Gelboin, H. V., and Gonzalez, F. J. Five of twelve forms of vaccinia virus-expressed human hepatic cytochrome P-450 metabolically activate aflatoxin B$_1$. *Proc. Natl. Acad. Sci. U.S.A.* 87, 4790, 1990.

33. Crespi, C. L., Penman, B. W., Steimel, D. T., Gelboin, H. V., and Gonzalez, F. J. The development of a human cell line stably expressing human CYP3A4: Role in the metabolic activation of aflatoxin B$_1$ and comparison to CYP1A2 and CYP2A3. *Carcinogenesis* 12, 355, 1991.

34. Yamano, S., Nhamburo, P. T., Aoyama, T., Meyer, U. A., Inaba, T., Kalow, W., Gelboin, H. V., McBride, O. W., and Gonzalez, F. J. cDNA cloning and sequence and cDNA-directed expression of human P-450 IIB1: Identification of a normal and two variant cDNAs derived from the CYP2B locus on chromosome 19 and differential expression of the IIB mRNAs in human liver. *Biochemistry* 28, 7340, 1989.

35. Gasser, R., Negishi, M., and Philpot, R. M. Primary structures of multiple forms of cytochrome P-450 isozyme 2 derived from rabbit pulmonary and hepatic cDNAs. *Mol. Pharmacol.* 32, 22, 1988.

36. Omiecinski, C. J. Tissue-specific expression of rat mRNAs homologous to cytochromes P-450b and P-450e. *Nucl. Acids Res.* 14, 1525, 1986.

37. Traber, P. G., Wang, W., McDonnell, M., and Gumucio, J. J. P-450IIB gene expression in rat small intestine: Cloning of intestinal P-450IIB1 mRNA using the polymerase chain reaction and transcriptional regulation of induction. *Mol. Pharmacol.* 37, 810, 1990.

38. Flammang, A. M., Aoyama, T., Gonzalez, F. J., and McCoy, G. D. Nicotine metabolism by cDNA-expressed human cytochrome P-450. *Biochem. Arch.* 8, 1, 1992.

39. Shimada, T., Misono, K. S., and Guengerich, F. P. Human liver microsomal cytochrome P-450 mephenytoin 4-hydroxylase, a prototype of genetic polymorphism in oxidative drug metabolism. *J. Biol. Chem.* 261, 909, 1986.

40. Gut, J., Meier, U. T., Catin, T., and Meyer, U. A. Mephenytoin-type polymorphism of drug oxidation: Purification and characterization of a human liver cytochrome P-450 isozyme catalyzing microsomal mephenytoin hydroxylation. *Biochem. Biophys. Acta* 884, 435, 1986.

41. Kawano, S., Kamataki, T., Yasumori, T., Yamazoe, Y., and Kato, R. Purification of human liver cytochrome P-450 catalyzing testosterone 6ß-hydroxylation. *J. Biochem.* 102, 493, 1987.

42. Ged, C., Umbenhauer, D. R., Bellew, T. M., Bork, R. W., Srivastava, P. K., Shinriki, N., Lloyd, R. S., and Guengerich, F. P. Characterization of cDNAs, mRNAs and proteins related to human liver microsomal cytochrome P-450 (S)-mephenytoin 4'-hydroxylase. *Biochemistry* 27, 6929, 1988.

43. Srivastava, P. K., Yun, S. H., Beaune, P., Ged, C., and Guengerich, F. P. Separation of human liver microsomal tolbutamide hydroxylase and (S)-mephenytoin 4'-hydroxylase cytochrome P-450 enzymes. *Mol. Pharmacol.* 40, 69, 1991.

44. Relling, M. V., Aoyama, T., Gonzalez, F. J., and Meyer, U. A. Tolbutamide and mephenytoin hydroxylation by human cytochrome P-450s in the CYP2C subfamily. *J. Pharmacol. Exp. Ther.* 252, 442, 1990.

45. Romkes, M., Faletto, M. B., Blaisdell, J. A., Raucy, J. L., and Goldstein, J. A. Cloning and expression of complementary DNAs for multiple members of the human cytochrome P-450IIC subfamily. *Biochemistry* 30, 3247, 1991.

46. Yasumori, T., Murayama, N., Yamazoe, Y., and Kato, R. Polymorphism in hydroxylation of mephenytoin and hexabarbital stereoisomers in relation to hepatic P-450 human-2. *Clin. Pharm. Ther.* 47, 313, 1989.

47. Rettie, A. E., Korzekwa, K., Kunze, K. L., Lawrence, R. F., Eddyll, A. C., Aoyama, T., Gelboin, H. V., Gonzalez, F. J., and Trager, W. F. Hydroxylation of warfarin by human liver cytochrome P-450: A role for proteins encoded by CYP2C9 in the etiology of anticoagulant drug interactions. *Chem. Res. Toxicol.* 5, 54, 1992.

48. Brolyn, F., Gaedigk, A., Heim, M., Eichelbaum, M., Mariko, K., and Meyer, U. A. Debrisoquine/sparteine hydroxylase genotype and phenotype: Analysis of common mutations and alleles of CYP2D6 in a European population. *DNA Cell Biol.* 10, 545, 1991.

49. Daly, A. K., Armstrong, M., Monkman, S. C., Idle, M. E., and Idle, J. R. The genetic and metabolic criteria for the assignment of debrisoquine 4-hydroxylation (cytochrome P-450IID6) phenotypes. *Pharmacogenetics* 1, 33, 1991.

50. Koop, D. R. and Casazza, J. P. Identification of ethanol-inducible P-450 isozyme 3a as the acetone and acetol monooxygenase of rabbit microsomes. *J. Biol. Chem.* 260, 1360, 1985.

51. Song, B. J., Veech, R. L., and Saenger, P. Cytochrome P-450IIE1 is elevated in lymphocytes from poorly controlled insulin-dependent diabetics. *J. Clin. Endocrinol. Metab.* 71, 1036, 1990.

52. Gonzalez, F. J., Ueno, T., Umeno, M., Song, B. J., Veech, R. L., and Gelboin, H. V. Microsomal ethanol oxidation system: Transcriptional and posttranscriptional regulation of cytochrome P-450. *CYP2E1.* Kalant, H., Khanna, J. M., and Israel, Y., Eds. *Alcohol & Alcoholism,* Supp. 1, 97–101, 1991.

53. Peter, R., Bocker, R. G., Beaune, P., Iwasaki, M., Guengerich, F. P., and Yang, C. S. Hydroxylation of chlorzoxazone as a specific probe for human liver cytochrome P-450IIE1. *Chem. Res. Toxicol.* 3, 566, 1990.

54. Nhamburo, P. T., Kimura, S., McBride, O.W., Kozak, C. A., Gelboin, H. V., and Gonzalez, F. J. The human CYP2F gene subfamily: Identification of a cDNA-encoding a new cytochrome P-450, cDNA-directed expression and chromosome mapping. *Biochemistry* 29, 5491, 1990.

55. Aoyama, T., Yamano, S., Waxman, D. J., Lapenson, D. P., Meyer, U. A., Fisher, V., Tyndale, R., Inaba, T., Kalow, W., Gelboin, H. V., and Gonzalez, F. J. Cytochrome P-450 hPCN3, a novel cytochrome P-450IIIA gene product that is differentially expressed in adult human liver. *J. Biol. Chem.* 264, 10388, 1989.

56. Kitada, M., Kamataki, T., Itahashi, K., Rikihisa, T., Kato, R., and Kanakubo, Y. Purification and properties of cytochrome P-450 from homogenates of human fetal livers. *Arch. Biochem. Biophys.* 241, 275, 1985.

57. Wrighton, S. A. and Vandenbranden, M. Isolation and characterization of human fetal liver cytochrome P-450HLP2: A third member of the P-450III gene family. *Arch. Biochem. Biophys.* 268, 144, 1989.

58. Komori, M., Nishio, K., Ohi, H., Kitada, M., and Kamataki, T. Molecular cloning and sequence analysis of cDNA containing the entire coding region for human fetal liver cytochrome P-450. *J. Biochem.* 185, 161, 1989.

59. Kitada, M., Kamataki, T., Itahashi, K., Rikihisa, T., and Kanakubo, Y. P-450 HFLa, a form of cytochrome P-450 purified from human fetal livers, is the 16α-hydroxylase of dehydroeprandrosterone 3-sulfate. *J. Biol. Chem.* 262, 13534, 1987.

60. Nhamburo, P. T., Gonzalez, F. J., McBride, O. W., Gelboin, H. V., and Kimura, S. Identification of a new P-450 expressed in human lung: Complete cDNA sequence, cDNA-directed expression, and chromosome mapping. *Biochemistry* 29, 5491, 1990.

61. Robertson, I.G.C., Philpot, R. M., Zeigler, E., and Wolf, C. R. Specificity of rabbit pulmonary cytochrome P-450 isozymes in the activation of several aromatic amines and aflatoxin B_1. *Mol. Pharmacol.* 20, 662, 1981.

62. Wolf, C. R., Statham, C. N., McMenamin, M. K., Bend, J. R., Boyd, M. R., and Philpot, R. M. The relationship between the catalytic activities of rabbit pulmonary cytochrome P-450 isozymes and the lung-specific toxicity of the furan derivative, 4-ipomeanol. *Mol. Pharmacol.* 22, 738, 1982.

63. Czerwinski, M., McLemore, T. L., Philpot, R. M., Nhamburo, P. T., Korzekwa, K., Gelboin, H. V., and Gonzalez, F. J. Metabolic activation of 4-ipomeanol by complimentary DNA-expressed human P-450: Species-specific metabolism. *Cancer Res.* 51, 4636, 1991.

64. Harris, C. C. Chemical and physical carcinogenesis: Advances and prospectives for the 1990s. *Cancer Res.* 51, 5023, 1991.

65. Sellers, T. A., Bailey-Wilson, J. E., Elston, R. C., Wilson, A. F., Elston, G. Z., Ooi, W. L., and Rothschild, H. Evidence for mendelian inheritance in the pathogenesis of lung cancer. *J. Natl. Cancer Inst.* 82, 1272, 1990.

66. Kellerman, G., Shaw, C. R., and Luyten-Kellerman, M. Aryl hydrocarbon hydroxylase inducibility in bronchogenic carcinoma. *N. Engl. J. Med.* 298, 934, 1973.

67. Paigen, B., Ward, E., Reilly, A., Houten, L., Gurtoo, H. L., Minowada, J., Steenland, K., Havens, M. B., and Sartori, P. Seasonal variation of aryl hydrocarbon hydroxylase activity in human lymphocytes. *Cancer Res.* 41, 2757, 1981.

68. Kouri, R. E., McKinney, C. E., Slomiany, D. J., Snodgrass, D. R., Wray, N. P., and McLemore, T. L. Positive correlation between high aryl hydrocarbon hydroxylase activity and primary lung cancer as analyzed in cryopreserved lymphocytes. *Cancer Res.* 42, 5030, 1982.

69. Kawajiri, K., Nakachi, K., Imai, K., and Hayashi, S.-I. The CYP1A1 gene and cancer susceptibility. *Crit. Rev. Oncol. Hematol.* 14, 77, 1993.

70. Caporaso, N., Landi, M. T., and Vineis, P. Relevance of metabolic polymorphisms to human carcinogenesis: Evaluation of epidemiologic evidence. *Pharmacogenetics* 1, 4, 1991.

71. Ayesh, R., Idle, J. R., Ritchie, J. C., Crothers, M. J., and Hetzel, M. R. Metabolic oxidation phenotypes as markers of lung cancer susceptibility. *Nature (London)* 312, 169, 1984.

72. Caporaso, N., Hayes, R. B., Dosemeci, M., Hoover, R., Ayesh, R., Hetzel, M., and Idle, J. R. Lung cancer risk occupational exposure and the debrisoquine metabolic phenotype. *Cancer Res.* 49, 3675, 1989.

73. Caporaso, N. E., Tucker, M. A., Hoover, R. N., Hayes, R. B., Pickle, L. W., Issaq, H. J., Muschik, G. M., Green-Gallo, L., Buivys, D., Aisner, S., Resau, J. H., Trump, B. F., Tollerud, D., Weston, A., and Harris, C. C. Lung cancer and the debrisoquine metabolic phenotype. *J. Natl. Cancer Inst.* 82, 1264, 1990.

74. Roots, I., Drakoulis, N., Ploch, M., Heinemeyer, G., Loddenkemper, R., Minks, T., Nitz, M., Otte, F., and Koch, M. Debrisoquine hydroxylation phenotype, acetylation phenotype, and ABO blood groups as genetic host factors of lung cancer risk. *Klin Wochenschr.* 66 (Suppl. XI), 87, 1988.

75. Speirs, C. J., Murray, S., Davies, D. S., Biola Mabadeje, A. F., and Boobis, A. R. Debrisoquine oxidation phenotype and susceptibility to lung cancer. *Br. J. Clin. Pharmacol.* 29, 101, 1990.

76. Caporaso, N. and Idle, J.R. The rationale for case-control methodology in epidemiological studies of cancer risk (response to Speirs et al., 1990). *Br. J. Clin. Pharmacol.* 30, 149, 1990.

77. Boobis, A. R. and Davies, D. S. Debrisoquine oxidation phenotype and susceptibility to lung cancer. *Br. J. Clin. Pharmacol.* 30, 653, 1990.

78. Crespi, C. L., Penman, B. W., Gelboin, H. V., and Gonzalez, F. J. A tobacco smoke-derived nitrosamine 4-(methylnitrosamino)-1-(3-pyridyl)-1-butanone, is activated by multiple human cytochrome P-450s including the polymorphic human cytochrome P-4502D6. *Carcinogenesis* 12, 1197, 1991.

Biomonitoring and Bioanalytical Chemistry

Important Considerations in the Ultratrace Measurement of Volatile Organic Compounds in Blood

David L. Ashley, Michael A. Bonin,
Frederick L. Cardinali, Joan M. McCraw,
Joe V. Wooten, and Larry L. Needham

INTRODUCTION

In the ultratrace analytical determination of organic compounds, investigators must consider certain factors, regardless of the nature of the analyte being measured or the matrix in which it is contained. These factors include sensitivity, selectivity, reproducibility, versatility, reliability, and sample stability.

The sensitivity of a method must be such that the analyte being examined can be measured at levels of biological significance. Often the levels of biological significance are not known until after the analytical system is developed and useful measurements can be made. If the detection limits are not low enough, a nondetectable result may occur and lead to a false conclusion of nonexposure.

The selectivity of a method is a critical consideration for the opposite reason. Interference resulting from poor selectivity suggests that an analyte is present when it is not. Neither of these situations is acceptable because the panic that may result from a false-positive exposure can be as detrimental as a false negative. Unfortunately these two analytical parameters tend to counter each other; adjusting instrumental parameters to increase sensitivity will often impair analytical selectivity and vice versa.

The reproducibility of an analytical system is extremely complex. There are many contributors to poor reproducibility, including instrument instability,

0-87371-951-4/95/$0.00+$.50

standard breakdown or contamination, and sampling system variation. The reproducibility of an analytical method often gets poorer as the method attempts to measure analytes at lower values. Thus, in ultratrace analysis, acceptable analytical reproducibility is difficult to achieve.

The versatility of a method frequently suffers as instrumental parameters are adjusted to maximize sensitivity and selectivity. The different properties of chemical species that allow one compound to be measured more efficiently with a particular method often decrease the efficiency with which another compound can be measured. Thus, the ability to measure individual compounds by a particular method will depend greatly on the compounds' similarity in certain critical properties that often are used to improve selectivity.

Sample stability can vary widely, depending on the stability of the analyte being examined, the state it is in when stored, and the conditions under which it is stored. Chemical instability, volatility, or reactivity, or the presence of reactive agents within the matrix, can each cause a sample to change abruptly after collection and before the analytical measurement is performed.

Finally, the reliability of a method to correctly answer the question that has been postulated also depends on investigators controlling the sources of contamination, collecting a representative sample, and correctly interpreting the results. All of these factors must be examined and their effects on the analytical results assessed before a method can be considered credible.

These factors are especially critical when trying to determine background levels of volatile organic compounds (VOCs) in blood. The properties of VOCs and their presence in residences, workplaces, and laboratories make them an extremely difficult group of compounds to measure at ultratrace levels. The volatility of VOCs can lead to a loss of material, which presents particular problems in sample collection, storage, and transfer and in the stability of the standard. Many VOCs are common solvents that are used in large amounts in analytical laboratories. Thus, when developing an analytical method, researchers must consider potential sources of contamination and ensure that such contamination is eliminated or at least minimized. Despite their ubiquity, extremely low levels of VOCs are found at background levels in blood. Thus, very good sensitivity is required if researchers are to assess these background levels. However, pushing detection limits to the extremes required to measure background levels of VOCs complicates efforts to achieve acceptable analytical reproducibility. VOCs are a large group of compounds with similar chemical properties, namely low molecular weight and low aqueous solubility. There are many compounds that fall into this category and they can provide complimentary information useful to researchers trying to pinpoint sources of exposure. Thus, it is highly beneficial to be able to assess a wide range of VOCs simultaneously. Unfortunately, methods that allow researchers to measure a wide range of VOCs in blood also cause compounds that are not of interest to be released from blood. Such release can be detrimental to analytical selectivity.

We have developed an analytical method that meets most of these factors. Using purge-and-trap/high-resolution gas chromatography/full-scan isotope dilution mass spectrometry, we can measure 31 VOCs with detection limits in the low parts-per-trillion (ppt) range in 10 mL of blood collected according to normal blood collection procedures. Complete details of this method can be found in a previous publication.[1]

MATERIALS AND METHODS

Safety

Many of the chemicals used in our method are considered extremely toxic, carcinogenic, or both. Safe operating procedures must be carefully designed and used whenever the neat materials are weighed, diluted, or manipulated in any way. All personnel involved in these procedures must be trained in the safe handling of toxic materials. Steps to prevent exposure include wearing personal protective equipment and using chemical fume hoods when handling materials. Blood samples, obtained directly from the field, were not screened for hepatitis B or the AIDS virus. For this reason, all samples are treated as infectious hazards. All sample handling was done within a biological safety cabinet and all items that came into contact with the blood were disinfected by using either a 10% sodium hypochlorite solution or an autoclave.

Materials

Unlabeled compounds were obtained from Aldrich Chemical Co. (Milwaukee, WI) or from Burdick and Jackson (Muskegon, MI). Deuterium- or carbon-13-labeled analogues were obtained (as the neat compounds) from either Cambridge Isotope Laboratories, Inc. (Woburn, MA) or Merck, Sharpe, and Dohme/ Isotopes (St. Louis, MO). Purge-and-trap-grade methanol (for dilution of standards) was also obtained from Burdick and Jackson.

Preparation of Materials

All glassware used in this study was carefully cleaned to remove any VOCs that may have been adsorbed onto the glass surface. We did this by washing glass items repeatedly with reagent-grade methanol, heating them at 150°C under vacuum (Lab-Line Instruments, Inc., Melrose Park, IL), and sealing them with Teflon-lined caps. Such cleaning was sufficient to prevent VOC contamination from the glassware. We obtained water (for use as a blank and for use in the final dilution of standards) from a rural well and distilled it while it was being purged with helium, thereby reducing the levels of VOCs below an acceptable level (<20 ppt). Neat analyte standards were first diluted in purge-and-trap-grade methanol to the intermediate stock solution concentrations our methods required. On the day of analysis these stock solutions were further diluted in blank water to achieve the desired final concentration. Isotopically labeled analogues were diluted from neat materials into purge-and-trap-grade methanol to produce intermediate stock solutions.

On the day of use, we further diluted the analogue solution. A 20-µL aliquot of this analogue solution was added to each blank, standard, unknown, or quality control (QC) sample before being injected into the purging apparatus. Positive-displacement pipettors (American Hospital Supply Co., Miami, FL) were used for all transfers of liquid in the microliter range. Quality control materials were prepared by combining serum obtained from multiple donors. This material was purged overnight with helium to reduce residual VOCs and then spiked to obtain the desired concentrations of individual VOCs. These samples were sealed in serum bottles (Wheaton, Millville, NJ) and frozen at –60°C until the day of use.

Serum was chosen as the quality control material instead of whole blood because of serum's greater homogeneity and long-term stability. Vacutainers containing potassium oxalate and sodium fluoride (Becton Dickinson Vacutainer Systems, Rutherford, NJ) were processed to remove VOC contamination that came chiefly from the rubber stopper. This process involved releasing the vacuum on each vacutainer, heating both tubes and rubber stoppers for 2 weeks at 70°C under vacuum, reassembling the vacutainers, restoring the vacuum, and sterilizing the vacutainers by using a cobalt-60 radiation source. Antifoam B emulsion (Sigma Chemical Co., St. Louis, MO) was dried either by exposure to room air or by heating it at 70°C in a vacuum oven. Between 10 and 50 mg of the dried emulsion was added to each sparging tube before it was attached to the purge-and-trap sampler. Before using, we cycled each sample location through a complete purging procedure to decontaminate the antifoam B and purging vessel.

Sample Preparation

Whole blood samples were collected by venipuncture into the processed vacutainers. These samples were carefully mixed and within 15 min were placed on wet ice or stored at refrigerator temperatures. Blood samples were removed from the vacutainers by puncturing the rubber stopper with a 1-in. 18-gauge needle attached to a 10-mL gastight syringe (Hamilton Co., Reno, NV). A 20-µL aliquot of the internal standard solution was added through the luer-lock of this syringe and the entire contents were injected into the sparging apparatus. The syringe was weighed before and after the sample was injected into the sparger to determine sample weight.

Instrumental Analysis

The purge-and-trap apparatus consisted of a Tekmar (Cincinnati, OH) LSC 2000 purge-and-trap concentrator with an attached ALS 2016 automated sampler. The helium flow rate was maintained at 30 mL/min at 20 psi. The purge-and-trap concentrator provided microprocessor control of all VOC isolation and

Table 10.1 Volatile Organic Compounds Measured in Human Blood

1,1,1-trichloroethane	1,1,2,2-tetrachloroethane	1,1,2-trichloroethane
1,1-dichloroethane	1,1-dichloroethene	1,2-dichlorobenzene
1,2-dichloroethane	1,2-dichloropropane	1,3-dichlorobenzene
1,4-dichlorobenzene	2-butanone	acetone
benzene	bromodichloromethane	bromoform
carbon tetrachloride	chlorobenzene	chloroform
cis-1,2-dichloroethene	dibromochloromethane	dibromomethane
ethylbenzene	hexachloroethane	m-/p-xylene
methylene chloride	o-xylene	styrene
tetrachloroethene	toluene	trans-1,2-dichloroethene
trichloroethene		

concentration steps including purge, dry purge, trapping on Tenax, and cryogenic trapping. The analytes were separated using a Hewlett-Packard (Avondale, PA) Model 5890 gas chromatograph with a heated interface to allow direct release of effluent into the mass spectrometer ion source. The chromatograph was equipped with a J & W (Folsom, CA) 30-m DB-624 column with a 1.8-μm film thickness.

The mass spectrometer was a VG Analytical (Manchester, UK) 70E high-resolution mass spectrometer operating at 3000 resolving power, with instrumental functions and data collection controlled by a 11-250J data system. Masses were calibrated vs. perfluorokerosene. The mass spectrometer was operated in full-scan mode (40–200 amu) with a scan rate of 1.0 sec/decade and a settling time of 0.2 sec. Table 10.1 shows the analytes measured using this method.

Because of the large number of area determinations required, automated processing was done by using either a separate 11-250J data system or a VAX 3100 workstation (Digital Equipment Corporation, Nashua, NH). Data were transferred directly from one system to another over an Ethernet network. Programming was developed for each of these data systems to provide target scan adjustment, accurate mass adjustment, and automated listing of results in a transmittable file. After adjusting for any deviation from accurate mass, we quantified the areas using a mass window of 0.03 amu.

Because of the coelution of styrene and o-xylene and identical mass fragments, the measured area values for styrene were corrected for the contribution from o-xylene as previously described.[1] Peak areas were transferred directly into an R:BASE (Microrim, Redmond, WA) data base developed specifically for this purpose. Then, to correct for the contribution of labeled analogues to the quantitation mass of the analyte and for the contribution of analytes to the quantitation mass of the labeled analogues, we used independently determined native and label mass ratios according to previously described methods.[2,3]

We made further statistical calculations and performed quality control steps using Statistical Analysis System software (SAS Institute, Inc., Cary, NC). We checked all runs for contamination (given by blank levels), acceptable instrumental operation (shown by internal standard area counts), and analyte sensitivity (indicated by reproducible quality control results).

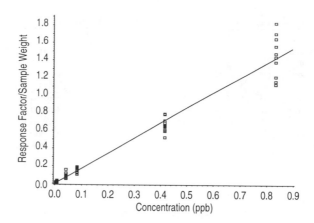

Figure 10.1. Standard curve (response factor vs. concentration) for tetrachloroethene with individual 95% prediction limits. R^2 = 0.987.

RESULTS AND DISCUSSION

A typical standard calibration curve for 1,1,1-trichloroethane is shown in Figure 10.1. Calibration for all analytes was performed using a six-point weighted linear fit of response factor divided by sample weight vs. concentration. Weighting factors were equal to the inverse of variance at individual standard concentration values, but in most cases weighting did not have a substantial effect on the slope and intercept calculated for this data. This curve shows excellent linearity over two orders of magnitude. In only a very few instances did we find an intercept significantly different from 0. Detection limits were calculated using previously described methods[4] and are based on the variation of calculated concentrations as the concentration approaches 0. The actual detection limits are equal to three times the standard deviation of the predicted concentration at 0 concentration. Detection limits for the VOCs determined by using this method have been given previously[1] and are in the low parts per trillion range for most of these analytes. The direct purging of blood and the use of magnetic sector mass spectrometry proved essential in achieving detection limits in this range.

Selectivity is an important consideration in the analysis of an untreated matrix. Blood is an extremely complex material, and analytical procedures incorporate various methods for separating the analyte of interest from interfering compounds. The purge-and-trap apparatus provides the first means of improving selectivity in addition to improving sensitivity by decreasing the relative concentration of some interferences. During the purging process, only volatile compounds are transferred into the analytical stream, and the adjustment of purging parameters provides some degree of variation in selectivity. Increasing the purging time or the sample temperature during purging yields a higher proportion of less volatile compounds. The second step in removing interferences involves the use of medium resolution mass spectrometry. An

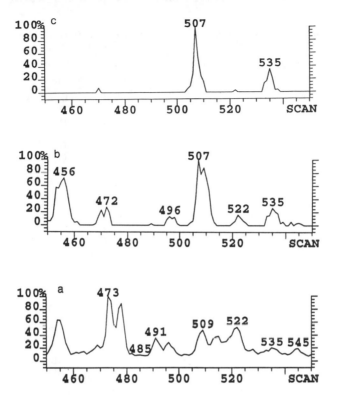

Figure 10.2. A typical whole blood sample demonstrating the improved chromatographic resolution possible with accurate mass using (a) total ion chromatogram, (b) nominal mass chromatogram (83 ± 0.5 amu), and (c) accurate mass chromatogram (82.94 ± 0.03 amu).

example of this improvement in selectivity from higher resolution is shown in Figure 10.2. Figure 10.2a, the total ion chromatogram between scans 450 and 550 for a typical blood sample, shows the expected result without any mass discrimination. Figure 10.2b shows the result when nominal mass discrimination is used. This chromatogram is a major improvement over that shown in 102a, since many of the peaks that are not of interest have been removed, but chloroform that is found at scan 507 overlaps another compound that also has a mass of 83 and elutes at scan 509. Figure 10.2c shows how the use of a ±0.03 amu mass window eliminates the interfering compound and provides an accurate measure of the level of chloroform in this sample. For many other analytes the removal of interferences was also absolutely necessary for correct quantitation.[5]

Eliminating sources of instrumental and sample contamination required additional efforts. Because of the low detection limits possible with this method, we are required to make special efforts to attain a blank water sample in order to test for instrumental contamination. In a previous paper[6] we showed that distilling water while it was purged with helium was an extremely effective way to remove VOC contamination. Except for chloroform, this procedure reduced

Table 10.2 Analysis of VOCs in Blank Water Stored in Commercial Vacutainers Compared with Blank Water Stored in Treated Vacutainers

Analyte	Detection limits[a]	Commercial vacutainers			Processed vacutainers		
		Mean[a]	(n)[b]	5–95%[a]	Mean[a]	(n)[b]	5–95%[a]
1,1,1-trichloroethane	0.045	0.26	(7)	0.10–0.46	ND[c]	(43)	ND–0.081
1,4-dichlorobenzene	0.042	ND	(7)	ND–0.012	ND	(42)	ND
benzene	0.026	ND	(7)	ND–0.054	ND	(42)	ND
bromoform	0.020	0.143	(7)	ND–0.27	ND	(43)	ND
chloroform	0.015	0.028	(7)	ND–0.061	ND	(41)	ND–0.045
ethylbenzene	0.025	0.029	(7)	ND–0.049	ND	(41)	ND
m-/p-xylene	0.034	0.073	(7)	ND–0.134	ND	(43)	ND
o-xylene	0.024	0.075	(7)	0.033–0.11	ND	(43)	ND–0.035
styrene	0.022	0.027	(7)	ND–0.064	ND	(43)	ND
tetrachloroethene	0.022	0.058	(7)	ND–0.079	ND	(42)	ND
toluene	0.059	0.10	(7)	ND–0.25	ND	(43)	ND–0.083
trichloroethene	0.007	0.022	(7)	ND–0.039	ND	(42)	ND

[a] In parts-per-billion (ppb).

[b] Number of samples.

[c] ND, Not detectable (result below detection limit).

all VOCs present in this mixture to levels below the limits of detection. In many cases, this was a reduction by more than a factor of 10. Blank water not only provides a daily measure of contamination within the instrument but also serves as the solvent for standard curves. This is critical, especially in the measurement of standards close to the detection limits when residual analyte levels in the diluent can prevent measurement at these low concentrations.

The materials used to collect and process samples must be analyzed as possible sources of contamination. In our study, since no sample workup was required prior to the analysis step, the vacutainers used for sample collection and storage are the primary sources of possible contamination. To test them, blank water was added to vacutainers which had not been processed and which had been specially treated to remove contamination. These were then stored under normal blood storage conditions for 1 to 2 weeks before analysis. Table 10.2 compares the levels of VOCs found in commercial vacutainers before and after treatment. Many of the VOC levels found in the vacutainers before cleanup were the same as or greater than the VOC levels measured in blood. In these cases, measured results are chiefly representative of vacutainer contamination and not indicative of blood levels. After cleanup, these vacutainers no longer showed measurable VOC contamination. Of particular note is bromoform, which is not normally found in the blood of persons who have not had specific exposure. Bromoform concentrations in blood collected in unprocessed vacutainers were in the high ppt range and detectable in all samples. These concentrations were a result of contamination from the vacutainers. Thus, VOC levels, when measured in blood collected in unprocessed vacutainers, are extremely suspect. We repeated our procedure with purged serum, and the qualitative results were the same even though the actual VOC contamination found in commercial vacutainers was increased because serum has a higher lipid content than water.

Table 10.3   **Results of Repeat Measurements of Spiked Serum Quality Control Materials**

Analyte	Detection limit[a]	Mean[a]	(n)[b]	Coefficient of variation (%)
1,1,1-trichloroethane	0.045	0.18	(47)	15.4
1,4-dichlorobenzene	0.042	0.76	(49)	11.1
benzene	0.026	0.12	(41)	10.7
chloroform	0.015	0.11	(47)	30.9
ethylbenzene	0.025	0.27	(50)	18.2
m-/p-xylene	0.034	1.24	(48)	7.1
o-xylene	0.024	0.94	(50)	13.0
styrene	0.022	0.15	(49)	13.6
tetrachloroethene	0.022	0.29	(48)	22.1
toluene	0.059	0.52	(47)	20.4
trichloroethene	0.007	0.064	(50)	26.4

[a] In parts-per-billion (ppb).
[b] Number of samples.

Table 10.4   **Results of Repeat Measurements of Normal Blood Examined over a 7-Week Period**

Analyte	Detection limit[a]	Mean[a]	(n)[b]	Coefficient of variation (%)
1,1,1-trichloroethane	0.045	0.078	(10)	22.4
1,4-dichlorobenzene	0.042	0.12	(9)	27.5
benzene	0.026	0.045	(10)	20.5
chloroform	0.015	0.029	(9)	25.2
m-/p-xylene	0.034	0.065	(10)	29.5
o-xylene	0.024	0.055	(9)	43.4
tetrachloroethene	0.022	0.048	(10)	22.0
toluene	0.059	0.20	(10)	25.2

[a] In parts-per-billion (ppb).
[b] Number of samples.

When characterizing method reproducibility, one must measure the analytes of interest at concentrations that are representative of the levels commonly seen in unknown samples. In a previous paper[1] we described repeat measurements of blood samples spiked at three different levels to cover the expected range of unknown levels. Relative standard deviations are usually less than 30%. As expected, most of the exceptions were found in the samples spiked with low levels of VOCs. These results actually showed a higher standard deviation than would be encountered in typical blood determinations, since they reflect variation in blood VOC levels both before and after the samples were spiked.

Table 10.3 shows that multiple measurements on spiked QC materials have lower standard deviations. Most of the VOCs showed coefficients of variation between 10 and 25%. This result is not surprising since the mean levels of this material are somewhat higher than those typically found in unspiked blood samples.

Table 10.4 shows another indicator of reproducibility of blood VOC measurements. Repeat measurements were made over 7 weeks on blood collected from an individual into 10 separate vacutainers. Not all analytes are included in this table because the levels of certain analytes in these blood samples were below the level of detection. The coefficients of variation for the analytes that were above the level of detection were generally in the 20–30% range.

The use of full scan mass spectrometry in place of the more commonly used technique, selective ion monitoring, provides increased versatility at the cost of poorer sensitivity. The use of selective ion monitoring increases sensitivity because only those particular ions of direct interest are monitored. The use of full scan measurement requires that analytical time is spent collecting data over mass ranges that are not of immediate interest. This drawback is outweighed, however, because full scan data can be reexamined later to gain analytical information about analytes that were not being considered when the sample was being examined. With purge-and-trap analysis of blood, this is especially advantageous because of the large number of additional analytes that enter the mass spectrometer ion source. In some cases, quantitative results can be determined by referencing these new analytes to labeled analogs that are already present. In other cases, qualitative statements are more appropriate. But whether qualitative or quantitative results are needed, the use of full scan mass spectrometry can give researchers a wealth of information that can be used to gain further insights into the presence of xenobiotics in human blood. We have already made use of this advantage by reexamining blood samples taken from fire fighters in Kuwait while the oil wells were burning. Using full scan mass spectrometry, we have identified and quantitatively measured a number of compounds in addition to those listed in Table 10.1.

Figure 10.3 shows repeat measurements of m-/p-xylene in blood from a single individual taken in unprocessed or treated vacutainers and stored at refrigerator temperatures for slightly longer than 7 weeks. The best-fit linear regression line is also drawn through these points to suggest the trend in results over time. Results of the VOC measurements of this person's blood suggest that the amount of m-/p-xylene in blood taken in prepared vacutainers does not change appreciably over 7 weeks when stored under these conditions. Results of measurements of the other detectable VOCs in the blood of this and other persons suggest that this is so for all of the VOCs reported here except for acetone and 2-butanone. The levels of concentration of these two analytes increased with time in storage. Since they are naturally occurring compounds, the increase in concentration is probably the result of continuing metabolism in the sample itself. Because of this increase in concentration, measurements of acetone and 2-butanone should be taken as soon as possible after collection. The levels of VOCs in samples collected in prepared tubes contrast distinctly with the levels of VOCs in samples stored in unprocessed commercial tubes (shown in Figure 10.3). For example, unprocessed vacutainers show a clear increase in m-/p-xylene concentration with storage. This is probably caused by continual leaching of VOCs from the vacutainer stopper into the sample. These same findings were consistent for many of the analytes examined in this study.

The reliability of an analytical method depends on the ability of those using it to make accurate, precise measurements at the concentrations of interest. The recovery of spiked VOCs using the method described here has already been

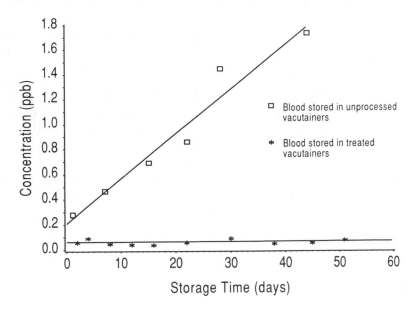

Figure 10.3. Change in blood concentration vs. storage time for *m-/p*-xylene in blood stored in unprocessed vacutainers and in treated vacutainers.

discussed.[1] For most individual spiking levels, the recovery is between 75 and 150%. These results are consistent across the entire range of added analyte, showing that the method is reliable not only at higher concentrations but also at the levels found in blood from nonoccupationally exposed persons. Most of the recovery results given in this paper are not substantially different from 100%, but there are some compounds that show high recovery rates. The exact reason for this finding is not clear, but it may be caused by specific matrix effects that are different for labeled and unlabeled analytes. Part of the recovery differences are undoubtedly caused by our inability to accurately assess the levels of VOCs in the unspiked blood that served as a baseline for calculating recovery.

CONCLUSIONS

We have developed a method that meets the conditions described at the beginning of this discourse. We were able to measure low parts-per-trillion levels of 32 VOCs in 10 mL of blood that were collected using normal blood collection procedures. These results are reproducible and reliable. The sensitivity of this method is better than that of previous methods because of the direct purging of the blood and the sensitivity enhancement that comes from the use of magnetic sector mass spectrometry. Selectivity was improved by using medium resolution mass spectrometry to differentiate among the many compounds extracted by the purge-and-trap technique. Versatility was enhanced by full scan measurements that permit a wide range of ions to be monitored and that allow for the assessment of analytes that may be important for future studies. Good reproducibility was accomplished by using isotope dilution mass spectrometry with labeled isotopes, which are available for virtually all of the compounds of interest. The VOCs reported here were stable in blood samples for at least 7 weeks when stored at refrigerator temperatures in specially prepared vacutainers. We prevented sample loss by using an air-tight syringe, rather than opening the vacutainers, thus ensuring a sample representative of the target. We confirmed the stability of the method by using QC samples run on each day and by carefully evaluating instrument parameters, blanks, QC materials, and standards. Sample collection was extremely critical, and the use of vacutainers that were specially processed to remove contamination was absolutely necessary to collect uncontaminated samples. By following these steps and being constantly vigilant for new sources of contamination and analyte loss, we have produced a reliable method for the measurement of VOCs in human blood at levels that are typically found in nonoccupationally exposed individuals.

ACKNOWLEDGMENTS

This research was funded in part by the Agency for Toxic Substances and Disease Registry. Use of trade names is for identification only and does not constitute endorsement by the Public Health Service or the U.S. Department of Health and Human Services.

REFERENCES

1. Ashley, D. L., Bonin, M. A., and Cardinali, F. L. et al. Determining volatile organic compounds in human blood from a large sample population using purge-and-trap gas chromatography/mass spectrometry. *Anal. Chem.* 64, 1021, 1992.
2. Cramer, P. H., Boggess, K. E., and Hosenfeld, J. M. Volatile organic compounds in whole blood-determination by heated dynamic headspace purge and trap isotope dilution GC/MS. Report No. EPA-560/5-87-008. U.S. Environmental Protection Agency, Washington, D.C., 1987.
3. Colby, B. N. and McCaman, M. W. A comparison of calculation procedures for isotope dilution determinations using gas chromatography mass spectrometry. *Biomed. Mass Spectr.* 6, 225, 1979.
4. Taylor, J. K. *Quality Assurance of Chemical Measurements.* Lewis, Chelsea, MI, 1987, pp. 79–82.
5. Bonin, M. A., Ashley, D. L., Cardinali, F. L., McCraw, J. M., and Patterson, D. G. Jr. Importance of high resolution in removal of contamination and interferences in measurement of volatile organic compounds in human blood by purge-and-trap/ gas chromatography/mass spectrometry. *J. Am. Soc. Mass Spectr.* 3, 831, 1992.
6. Cardinali, F. L., McCraw, J. M., Ashley, D. L., and Bonin, M. A. Production of blank water for the analysis of volatile organic compounds in human blood at the low parts-per-trillion level. *J. Chromatogr. Sci.* 32, 41, 1994.

CHAPTER 11

Applications of New HPLC/MS Techniques in Human Biomonitoring

William M. Draper, F. Reber Brown, Robert A. Bethem,
Michael J. Miille, and Charles E. Becker

INTRODUCTION

Biomarkers in Epidemiology and Industrial Hygiene

Hulka and Wilcosky[1] have discussed the advantages of using biological markers in epidemiologic research. They review a classification scheme first introduced by Perera and Weinstein[2] that identifies five biological marker categories: internal dose, biological effective dose, biological response, disease, and susceptibility. These markers delineate the sequence of events between chemical exposure and the development of exposure-related disease.

Internal dose biomarkers are widely used, as evidenced by the research presented in this book, and provide a measure of the integrated exposure to a chemical insult from all exposure routes. Internal dose measurement sometimes relies on determination of the unchanged chemical in excreta (i.e., exhaled solvents), but most often metabolic products in the blood and urine are monitored. The degree to which a toxicant penetrates and interacts with critical subcellular receptors is indicated by biological effective dose markers (i.e., DNA or protein adducts). Cells and tissues respond with biochemical changes including those to chromosomes (i.e., sister chromatid exchange) or modification of the activity of specific enzymes (i.e., acetylcholinesterase)—biochemical changes are monitored with biological response markers. Last, and most important, the onset of disease is manifested prior to the appearance of clinical symptoms by biochemical and biological events associated with disease markers. Disease markers are not only important tools in epidemiology, but also are essential to many disease screening programs.

Internal dose biomarkers are used in industrial hygiene as well. The American Conference of Governmental Industrial Hygienists (ACGIH) publishes biological exposure indices (BEIs) for use in assessing occupational exposure to chemicals.[3] The BEI is a reference value indicating typical concentrations of internal dose biomarkers, so-called determinants, in urine, blood, or exhaled air for inhalation exposure of healthy workers at the threshold limit value (TLV). BEIs have been published for only about 20 synthetic organic chemicals, mostly solvents, but ACGIH has a standing committee to study and establish BEIs to new chemicals. Urine is the most frequently used biological medium, and urinary metabolites are determinants in all but a few cases.

LC/MS in Human Biomonitoring

Liquid chromatography/mass spectrometry (LC/MS) combines the ability to separate complex mixtures of involatile substances with a spectroscopic technique of high sensitivity and specificity. High-pressure liquid chromatography was introduced in the early 1970s and its versatility was evident almost immediately. HPLC solved many separation problems encountered with polar, involatile, and thermally unstable substances (PITS). Prior to HPLC, polarity and involatility were addressed by derivatization with alkylating, acylating, or silylating reagents, which are hazardous to work with, but render the analytes sufficiently volatile for gas chromatography. HPLC, however, has always been limited by the lack of a universal, sensitive, and specific detector, a problem compounded by the relatively low chromatographic efficiency of HPLC columns. Therefore, each advancement in LC/MS technology and new LC/MS design has been greeted by an eager audience of analytical chemists.

There were many engineering obstacles to overcome in linking mass spectrometers to liquid chromatographs as outlined in a recent review article by Cairns and Siegmund.[4] The most fundamental limitation is that mass spectrometer sources and analyzers operate at a high vacuum (e.g., 10^{-5} torr) that is difficult to maintain when condensed HPLC mobile phases are introduced. For example, 2 mL/min of acetonitrile equates to a gas flow rate of 1.1 L/min at 25°C and 1 atm.

Over the years various LC/MS designs have been reported including direct liquid introduction (DLI), moving belt interface (MBI), thermospray (TSP), and particle beam (PB),[4] and each of these techniques has proven useful in analyzing the PITS. TSP was used to determine carbamates,[5] dyes,[6] organophosphorus (OP) insecticides,[7,8] and triazine herbicides[8] and DLI LC/MS was useful for determination of aldicarb,[9] benzimidazoles,[10] and carbamate[11] and OP[12] insecticides. PB mass spectrometry is too young to have established much of a track record, but recent reports describe its use in the determination of daminozide,[13] ethylenethiourea,[14] chlorophenoxy,[15] and phenylurea herbicides[16] and benzidines.[17]

LC/MS is now routinely used in industry to investigate mammalian metabolism of veterinary drugs, pharmaceuticals, and pesticides. TSP LC/MS has been applied to the quantification of metabolites of chlorambucil,[18] mycotoxins,[19,20] nicotine,[21] trimethyllead,[22] and warfarin[23] and many other compounds. The metabolic products of benzo[a]pyrene,[24] benzodiazepines,[25] and 2,4,6-trinitrotoluene[26] have been determined by DLI LC/MS.

The growing body of literature on LC/MS applications in metabolism research suggests that LC/MS may be useful in biological monitoring. Each of the metabolites elaborated is a potential internal dose biomarker. What remains to be fully demonstrated is that LC/MS techniques can make the transition from qualitative or semiquantitative procedures to those useful for routine quantitative analysis. The expanding use of LC/MS in state-of-the-art drug testing laboratories[27,28] argues that this is indeed the case.

Various practical considerations such as cost, reliability, operator training requirements, and sample throughput bear upon the selection of analytic techniques for use in human biomonitoring. As a result, gas and liquid chromatography, and more recently GC/MS have established themselves as the workhorse techniques in human biomonitoring. The goal of this chapter is to provide some insights into the unique capabilities of liquid chromatography/mass spectrometry that may be useful in human biomonitoring. Specifically, we will consider two popular LC/MS interfaces, thermospray and particle beam, including a brief discussion of their operating principles and performance characteristics. A further objective will be to summarize findings from a multiyear study of LC/MS biomonitoring applications in our laboratory and to note some of the relevant literature.

LIQUID CHROMATOGRAPHY OF BIOMARKERS

One of the earliest conclusions of our work was that the conventional reversed-phase HPLC would be inadequate for many useful biomarkers. The mammalian metabolism of benzene illustrates the diverse pathways involved and the polarity of products elaborated (Fig. 11.1). The predominant excretion products of most xenobiotics are in fact glucuronide and sulfate conjugates. In the case of benzene, the sulfuric acid conjugate is most abundant and there are many minor metabolites including highly functionalized phenolic compounds, a dicarboxylic acid, and mercapturates (Fig. 11.1). Without the use of paired ion chromatography (PIC) reagents, most of these compounds cannot be sufficiently retained on a reversed-phase HPLC column.

As the important urinary metabolites are usually acidic, strong anion-exchange (SAX) chromatography has been found to be useful. We first used silica-based SAX columns as a general separation method for glucuronides, sulfates, and phenols.[29] Styrene-divinyl benzene resin SAX columns, however, tolerate a much broader pH range and are more durable.[30] A further advantage of the resin-based columns is that reversed-phase interactions contribute to chromatographic retention. Accordingly, the major urinary metabolites of benzene as well as the parent compound are each separated on a resin-based SAX column

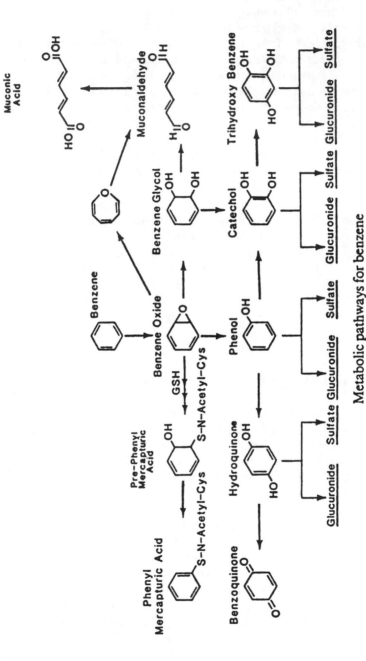

Figure 11.1   Metabolic pathways of benzene in mammals.

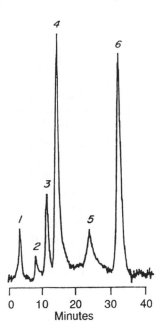

Figure 11.2   PB LC/MS total ion current chromatogram showing strong anion-exchange separation of benzene metabolites: phenol (1), phenylglucuronide (2), phenylsulfate (3), phenylmercapturic acid (4), benzene (5), and *trans,trans*-muconic acid (6). Compounds were eluted from a 4.6 mm × 25 cm Hamilton PRP-X100 column with 0.25 mL/min pH 5.3 0.15 *N* ammonium formate–acetonitrile (33:67, v/v). (Reproduced from the *Journal of Chromatographic Science* by permission of Preston Publications, a Division of Preston Industries, Inc.[30])

(Fig. 11.2). The retention of the metabolites on these columns can be modified by selection of buffer, ionic strength, and pH as well as the type and amount of mobile phase modifier.[30]

Both the composition and flow rate of the mobile phase must be compatible with the specific LC/MS interface used, as will become clear later. TSP performance is optimal with a flow rate of 1.0–2.0 mL/min using a high-water-content mobile phase—PB flow rates are limited to about 0.2–0.6 mL/min and low water content is optimal. Accordingly, 4.6-mm (i.d.) columns are best suited to TSP and 2.1-mm (i.d.) columns are widely used in PB LC/MS.

Suspended particulate matter in samples, solvents, and buffers must be removed by ultrafiltration. Volatile buffers such as ammonium acetate are usually required in TSP MS and are widely used in PB as discussed in subsequent sections. Here again the strong anion-exchange separations described are well suited as anions are eluted with ammonium acetate or formate buffer. Involatile salts, buffers, and PIC reagents are the bane of LC/MS because the buildup of deposits in TSP probes, PB momentum separator cones, and the source components can rapidly degrade instrument performance.

THERMOSPRAY LC/MS

The TSP LC/MS Interface and TSP Mass Spectrometry

The discovery of thermospray ionization was a major advancement in LC/MS research and development. Blakely and co-workers[31] found that ionized species were produced by involatile analytes directly from liquids without external ionizers. In the TSP instrument the HPLC column eluent is conveyed directly to a heated metal capillary where over 90% of the mobile phase volatilizes—a jet of vapor and aerosol exits the vaporizer probe into an evacuated chamber. Involatile species (about 80%) are enriched in the aerosol phase. On entering the heated source block, ions are ejected and extracted via a sampling cone to the mass spectrometer analyzer. Although optional under some circumstances, the source is equipped with a heated filament for "filament-on" TSP ionization. The large volume of vapor generated on volatilization of the mobile phase is removed by a high capacity mechanical pump and cold trap.

An important feature of TSP ionization is that the analyte need never volatilize from the liquid phase prior to ionization. As such, TSP is classified as a desorption ionization technique[32] and provides spectra similar to those obtained by field desorption ionization mass spectrometry.[31] A volatile buffer (e.g., ammonium acetate or formate) is needed to ionize most analytes in a process known as "buffer ionization." Therefore, an important constraint of TSP LC/MS is that the chromatographic mobile phase almost always must contain a volatile buffer. There are some exceptions as in the negative ion TSP determination of substituted phenols described later.

Shortly after the discovery of TSP ionization, and the appearance of commercial TSP instruments, exciting applications in biology and biochemistry were reported in the literature. Notably, TSP LC/MS provided spectra of glucuronide[33] and sulfate conjugates,[34] acetylcholine,[35] and the metabolites of L-carnitine in biological fluids without elaborate sample preparation or derivatization.[36] Obtaining mass spectra for these compound classes would be impossible by GC/MS or even by sample introduction using a heated solid probe.

The TSP mass spectrum of hippuric acid is shown in Figure 11.3. The positive ions observed with 0.1 M ammonium acetate in the mobile phase are the protonated (MH^+, m/z 180) and ammoniated molecule ions (MNH_4^+, m/z 197). The deprotonated molecule ($[M - H]^-$) and acetate adduct ($[M + acetate]^-$) ions are abundant in the negative ion TSP spectrum. Similarly, phenyl glucuronide analyzed with filament on and a mobile phase of methanol–0.1 M ammonium acetate (1:1, v/v) gives an m/z 288 (MNH_4^+) base peak and two additional ions, m/z 302 ($[M + CH_3OH]^+$ or more likely $[M + CH_3OH - H_2O + NH_4]^+$, 57%) and m/z 194 (ammoniated dehydroglucuronic acid, 17%). The volatile buffer and mobile phase organic modifier are important as different adduct ions and clusters will result. When using methylene chloride in normal-phase TSP LC/MS,

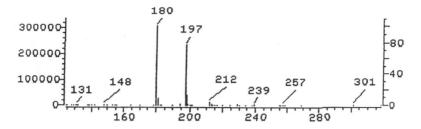

Figure 11.3   Thermospray spectrum of hippuric acid: m/z 180, MH$^+$; m/z 197, MNH$_4^+$.

for example, a high degree of chloride attachment giving [M + Cl]$^-$ ions is observed.[37]

Very little fragmentation is observed in the hippuric acid spectrum, which is typical of TSP and other "soft ionization" methods. In TSP ionization very little excess energy is imparted to the analyte and fragmentation is minimal. TSP ionization has received considerable attention in the literature: TSP spectra are strongly influenced by gas-phase ion/molecule equilibria, and ion attachment vs. protonation/deprotonation can be predicted on the basis of gas-phase acidities.[38] Analyte ions also form clusters with ammonia, acetic acid, and solvent molecules, the intensities of which are determined by additional gas-phase equilibria.[39] The pH of the mobile phase has little influence on TSP mass spectra.[40]

The complexity of TSP ionization explains several of the limitations of TSP mass spectrometry. Ion intensities are affected by flow rate,[31] and TSP ionization efficiency and fragmentation patterns tend to vary from day to day. Furthermore, TSP spectra are influenced by pressure, temperature, buffer concentration, vapor composition, sample type, and instrument design.[41,42] The spectrum of phenyl glucuronide obtained under similar conditions on a different instrument, for example, had m/z 288 and m/z 194 ions, but lacked the methanol cluster ion. Because of the variability of TSP spectra and the limited number of ions present, TSP mass spectral libraries are not widely used.

TSP LC/MS Determination of Substituted Phenols

Substituted phenols are important urinary metabolites of many industrial compounds including herbicides, insecticides, disinfectants, and wood preservatives. Chlorophenols also are produced in the chlorination of drinking water. With their many uses and wide distribution in the environment, chlorinated phenols are often found in biological samples.[43] Urinalysis has been used to determine occupational and extraneous exposures to pentachlorophenol (PCP) and other chlorophenols.[44,45] Our laboratory is frequently called on to determine PCP residues in wastes and water as well as human blood and urine. Phenols are usually derivatized with diazomethane,[46] pentafluorobenzyl bromide,[46] or anhydrides before analysis by gas chromatography, the conventional analytical method.

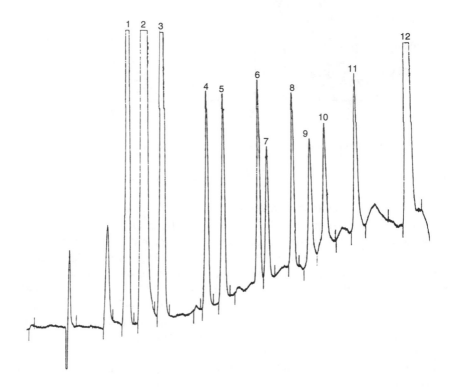

Figure 11.4   HPLC separation of substituted phenols. An RP-8 reversed-phase column (Spheri-5, 5 μm, 250 × 2.1 mm, Brownlee Labs, Santa Clara, CA) was used with a flow rate of 0.5 mL/min and a 30 min gradient from 40 to 70% methanol in water (the methanol contained 0.05% acetic acid). The following phenols were detected by UV absorbance: phenol (1), 2,4-dinitrophenol (2), 4-nitrophenol (3), 4-chlorophenol (4), 4-bromophenol (5), 4-chloro-2-methylphenol (6), 2,4-dichlorophenol (7), 2,4-dibromophenol (8), 2,4,6-trichlorophenol (9), 2,4,5-trichlorophenol (10), 2,3,4,6-tetrachlorophenol (11), and pentachlorophenol (12).

With TSP LC/MS, the direct determination of substituted phenols including PCP is possible.

The separation of phenols on a C_8 reversed-phase column using a water–methanol gradient and LC/UV detection is shown in Figure 11.4. The organic modifier contains 0.05% acetic acid to suppress ionization, improve peak shape, and increase retention. In negative ion TSP, the phenols are detected as molecular anions or deprotonated molecule ions. With a filament to induce ionization a number of unusual ions are detected including [M + H]⁻ and [M + Cl]⁻ as seen in the spectrum of trichlorophenol (Fig. 11.5). These ions also have been reported by Jones and co-workers in their study of chlorophenoxy acid herbicides.[47] TSP spectra are unique depending on whether ionization is induced by discharge electrode or filament.

The single ion traces obtained for a mixed phenol standard are shown in Figure 11.6. Using single ion monitoring (SIM), the instrument detection limit

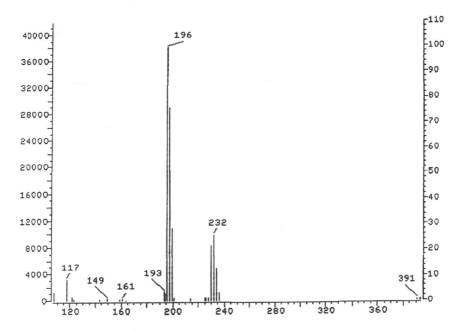

Figure 11.5   Negative ion thermospray spectrum of 2,4,5-trichlorophenol. filament on; flow injection in methanol–water (1:1, v/v); m/z 196, M⁻; m/z 230, [M – H + Cl]⁻.

for PCP was about 10 ng on column (S/N = 13), not as sensitive as an electron capture GC detector, but nevertheless exhibiting far greater specificity. Wright and co-workers previously used DLI LC/MS for direct determination of chlorophenols in human urine.[48] Their detection limits for each of 19 chlorophenols ranges from 2 to 10 ng.

There have been several reported applications of TSP LC/MS in human biomonitoring, one involving estimation of cigarette smoke exposure and a second in which tetraalkyllead exposures were determined. McManus and co-workers quantified trans-3′-hydroxycotinine, cotinine, demethylcotinine, nicotine, and 14 additional nicotine metabolites in urine samples from cigarette smokers.[21] These investigators separated the metabolites by reversed-phase HPLC, introduced ammonium acetate postcolumn, and quantified the materials against a trideuterated cotinine internal standard.

Tetraalkyllead is metabolized in the liver and undergoes excretion in the kidney as trialkyllead cation and other products. Niedhart and co-workers[49] determined trimethyllead ion in human urine by TSP LC/MS using a reversed-phase column and a syringe pump and by monitoring the m/z 253 ion, $(CH_3)_3Pb^+$. Their detection limit was 5 ng and TSP MS detector response was linear between 5 and 500 ng.

TSP/Tandem Mass Spectrometry

The lack of fragmentation in TSP mass spectrometry is advantageous in obtaining molecular weights, but limits the amount of structural information

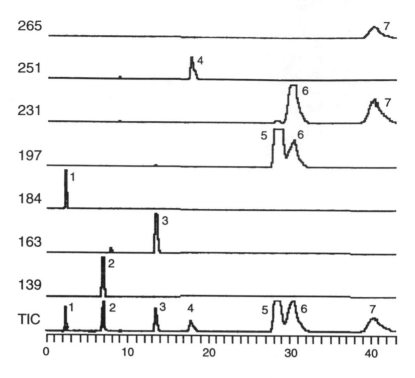

Figure 11.6   Negative ion thermospray LC/MS (filament on) separation of substituted phenols. A 4.6 × 100 mm RP-8 column was eluted with 1.0 mL/min of methanol–water–acetic acid (1:1:0.005, v/v/v) and compounds were detected by SIM as follows: 2,4-dinitrophenol (*m/z* 184), 4-nitrophenol (*m/z* 139), 2,4-dichlorophenol (*m/z* 163), 2,4-dibromophenol (*m/z* 251), 2,4,5-trichlorophenol (*m/z* 197), 2,3,5,6-tetrachlorophenol (*m/z* 231), and PCP (*m/z* 265).

and the number of ions available for selected ion monitoring (SIM). Minor TSP source modifications including filaments, discharge electrodes, and repellers have been used to promote fragmentation and enhance sensitivity. By far the most powerful adaptation has been the use of tandem mass spectrometers. Molecular ions formed in the TSP source serve as precursor ions which pass from the first quadrupole (Q_1) to the second quadrupole (Q_2), where they are accelerated and collide with a collision gas, usually argon. Control of the accelerating voltage and the collision gas pressure determine the severity of the collisions and the extent of collision activated dissociation (CAD or CID) fragmentation. CAD product ions are then analyzed in a third quadrupole (Q_3) and displayed as the product ion spectrum.

When the ammonium adduct ion of the phenyl glucuronide (MNH_4^+, *m/z* 288) is subjected to CAD fragmentation, the product ions include a prominent aglycone MH^+ ion and a suite of ions from the glucuronic acid fragment, *m/z* 113, *m/z* 159, and *m/z* 193 or 194.[50] The glucuronic acid-derived product ions form a fingerprint that is diagnostic of all glucuronide conjugates. CAD fragmentation of aryl sulfuric acid conjugates is very simple in comparison with only neutral loss of SO_3 ([M − 80]⁻) being observed.[51]

TSP MS/MS Determination of Aromatic Amines

We examined the use of MS/MS in the determination of aromatic amines. The compounds studied are known or suspected carcinogens regulated in California as mandated by the Safe Drinking Water and Toxic Enforcement Act of 1986, so-called Proposition 65. Aromatic amines and their derivatives also are urinary metabolites of various synthetic dyes, and have been used as internal dose biomarkers.[52,53]

A total ion current chromatogram showing 9 aromatic amines and derivatives is depicted in Figure 11.7. Each of the amines produced the expected protonated molecule ion (MH$^+$) as a base peak, but only one additional ion, an acetonitrile cluster ion ([M + ACN + H]$^+$), was routinely observed. Diphenylhydrazine produced an unexpected base peak of m/z 183 indicating that this reactive amine may be oxidized to azobenzene in the TSP source.

The complexity of CAD product ion spectra contrast sharply with the TSP spectra. The MH$^+$ ion of m-cresidine (m/z 138), for example, fragmented in various ways with an acceleration voltage of 15 eV and an argon pressure of 10 mtorr. The fragmentation pathways result from the loss of various neutral radicals and small molecules including hydrogen atoms, alkoxy and alkyl radicals, ammonia, formaldehyde, and methanol (Fig. 11.8).

Selected Reaction Monitoring

A discussion of tandem mass spectrometry, regardless of how brief, would be incomplete without some mention of selected reaction monitoring (SRM). When the scanning of both Q_1 and Q_3 quadrupoles is linked, analytes undergoing loss of a common neutral fragment are detected selectively. Selective detection of sulfuric acid metabolites is thus accomplished by passing ions of mass M at Q_1 and mass M – 80 (M – SO$_3$) at Q_3.[51] Similarly, glucuronides and glutathione conjugates are detected by neutral loss scanning of M – 176[51,52] and M – 129,[54,55] respectively. Sulfate conjugates are precursors of the m/z 97 product ion.[54] Each stage of mass filtering results in transmission losses meaning that the absolute sensitivity drops in tandem mass spectrometry. Useful sensitivity may increase, however, as the signal-to-noise ratio is improved. The applications of SRM in the determination of glucuronides, sulfates, mercapturates, carboxylates, and other important metabolites have been reviewed by Straub and co-workers.[54,56]

PARTICLE BEAM LC/MS

The PB Interface

The particle beam interface is a recent development in LC/MS technology having been first described by Willoughby and Browner in 1984.[57] In PB LC/MS, a conventional electron ionization (EI) MS source is used and conventional EI

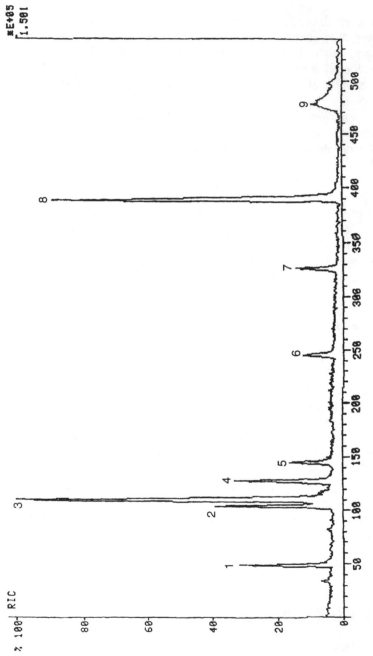

Figure 11.7   Positive ion thermospray LC/MS of aromatic amines. A C_{18} reversed-phase column (3 μm, 4.6 × 150 mm) was eluted with a 0.1 M pH 4.5 ammonium acetate–acetonitrile gradient—the acetonitrile content was held at 30% for 1 min and then increased to 70% in a 13 min interval. The compounds determined were 4-methoxy-1,3-phenylenediamine (1), m-cresidine (2), 4-aminophenylether (3), o-anisidine (4), o-toluidine (5), 2-aminonaphthalene (6), diphenylhydrazine (7), p,p-bis(dimethylamino)benzophenone (8), and auramine (9).

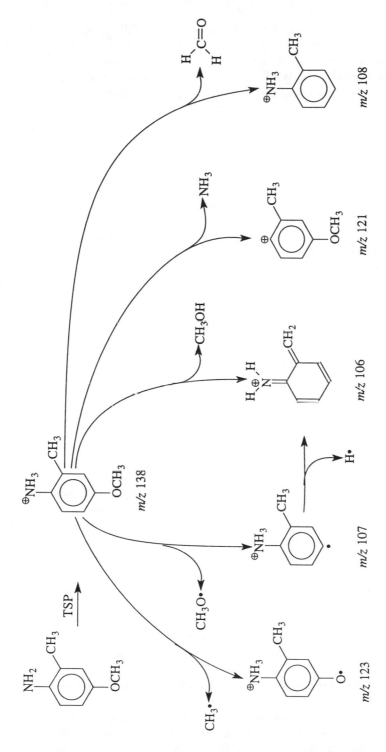

Figure 11.8 CAD fragmentation of the *m*-cresidine TSP molecule ion, *m/z* 138. The collision energy was 15 eV and the argon collision gas pressure was 10 mtorr.

spectra are obtained. This is not true of either TSP or DLI LC/MS. PB therefore refers strictly to an interface for sample introduction, while TSP is both a sample introduction device and an ionization technique.

Browner and co-workers[56] provide a detailed discussion of the PB interface. Briefly, the interface disperses the incoming HPLC eluent either by use of a pneumatic nebulizer or by rapid heating. The aerosol, entrained in a stream of helium, is heated in a desolvation chamber maintained at ~45°C and at a pressure of ~0.3 atm. Higher desolvation chamber temperatures are useful for mobile phases with a high water content. The aerosol particles grow smaller in size in the desolvation chamber as the mobile phase volatilizes. The aerosol–vapor mixture then passes through a nozzle where supersonic expansion occurs. Enrichment is achieved by skimming the expansion jet in a two-stage momentum separator—particles with the greatest momentum have the highest transport efficiency. Sample components not removed in the momentum separator are transferred via a heated tube to the ion volume of an MS source where flash vaporization takes place.

Unlike TSP, samples must possess some volatility for ionization and detection. In the PB LC/MS source molecules are ionized by bombardment with electrons released from a heated filament and accelerated in an external field. The extent of fragmentation is altered by changing the electron acceleration voltage from the nominal 70 eV, which gives acceptable fragmentation and sensitivity for most analytes. Lowering the acceleration voltage "softens" the ionization and enhances the intensity of the molecular ion. Alternatively, introduction of a reagent gas (e.g., methane, isobutane, or ammonia) allows chemical ionization (CI) in PB LC/MS, another soft ionization technique that enhances molecular ions. Both CI PB LC/MS and TSP can provide information complimentary to EI PB LC/MS.

There are a number of advantages to employing electron ionization in LC/MS. First, in EI mass spectrometry there is considerable fragmentation of molecular ions, which increases both the amount of structural information and the number of ions available for SIM. Second, and more important is that electron ionization is much more reproducible than TSP. EI mass spectra recorded on one instrument, whether quadrupole or magnetic sector, have the same fragment ions and, to a certain extent, ion abundances as those found with other EI mass spectrometers. As such, the spectra are searchable. That is, by comparing the spectrum of an unknown with a library of reference spectra, tentative identification is possible. Modern EI mass spectrum libraries contain many thousands of entries—for example, the Probability-Based Matching system contains 41,429 spectra of 32,403 different compounds.[59] A computerized search can be accomplished in a few seconds using modern MS data systems.

Identifying Unknown Metabolites by Library Search in PB LC/MS

The identification of unknown substances by comparison of sample spectra with a "library" of EI spectra is widely used in the enforcement of drinking water standards, hazardous waste regulations, and other types of regulatory monitoring. The importance of this capability in human biomonitoring is easily appreciated. We evaluated the "identification power" of a PB mass spectrometer by challenging the instrument with polar metabolites of important pesticides. Ten micrograms of each compound (10 µL of µg/µL in methanol or water) was introduced by flow injection (no column) with a mobile phase of 0.5 mL acetonitrile/min—spectra generated were then subjected to a computerized library search.

The PB MS spectrum of 3-hydroxycarbofuran (3-HC), a polar carbofuran metabolite, is shown in Figure 11.9A. A weak M^+ ion is present (m/z 237) as well as two fragment ions due to loss of methylisocyanate (55%, m/z 180, $[C_{10}H_{11}O_3]^+$) and methylisocyanate followed by isopropane (base, m/z 137, $[C_7H_5O_3]^+$). Another carbofuran metabolite, 3-ketocarbofuran phenol, gave the M^+ as base peak (m/z 178) and intense m/z 137 ([M – propene]$^+$) and m/z 163 fragment ions ([M – CH_3]$^+$) (Fig. 11.9E). The PB mass spectrum of 3-HC closely matched the library EI spectrum (Fig. 11.9B) allowing the data system to correctly identify the metabolite.

The PB EI spectrum of diethylphosphoric acid is very similar to the EI library spectrum, again allowing correct structure assignment (Fig. 11.9C and D). In either case an MH^+ ion (m/z 155) occurs in place of the expected M^+. Formation of the protonated molecule ion under PB MS conditions may be due to decomposition in the source, or possibly CI. The library spectrum may have been obtained by solid probe sample introduction where similar decomposition may have occurred. The fragment ions and their relative intensities (i.e., m/z 99, base; m/z 127, 53%, $[M – C_2H_5]^+$; m/z 109, 45%, $[M – OCH_2CH_3]^+$) are in good agreement. Diethylphosphate is one of several alkylphosphates used as internal dose biomarkers of commercial OP insecticides.[60] Alkylphosphate determination by GC-based methods has been challenging to agricultural chemists for many years,[61] and LC/MS may prove a useful alternative.

In a recent PB LC/MS study of styrene metabolism in humans, both mandelic and hippuric acids were identified by computerized library search.[62] Another styrene metabolite, phenylglyoxylic acid, had no record in the National Institute of Standards and Technology (NIST) library, but was matched to a structurally related compound, 1-phenyl-1,2-propanedione.

Several points should be emphasized in regard to the search capability of PB LC/MS. First, there are no records for most metabolite structures in existing spectral libraries. Accordingly, users will find it beneficial to create their own library of interesting compounds. Second, while PB LC/MS is capable of identifying metabolites that cannot be directly identified by GC/MS, the problem remains for the chemist to identify the substance from which the metabolites were elaborated, and to which the subject was exposed. For example, mandelic, phenylglyoxylic, and, to a lesser extent, hippuric acid are well known determinants of styrene. Finally, the identification of compounds by computerized library search is tentative. Some MS purists claim that library searching can only be regarded as an aid to mass spectrum interpretation. The spectrum of 3-ketocarbofuran phenol (Fig. 11.9E), for example, has similar spectral features to 3-hydroxybenzaldehyde oxime (Fig. 11.9F), and the MS data system faith-

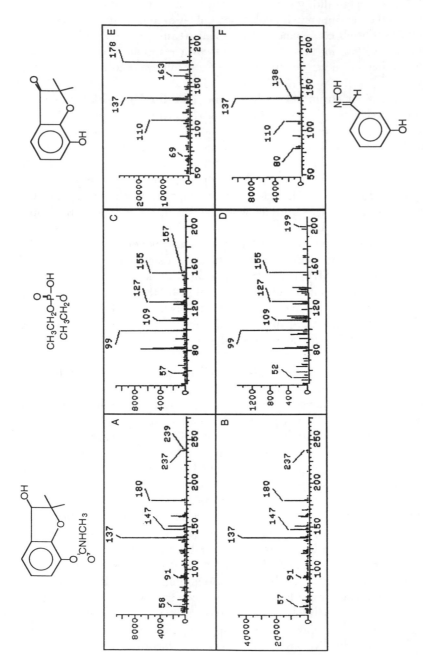

Figure 11.9   Particle beam LC/MS spectra of 3-hydroxycarbofuran (A), diethylphosphoric acid (C), and 3-ketocarbofuran phenol (E). The corresponding NIST EI library spectra (B,D,F) obtained by computerized search—3-ketocarbofuran had no NIST record and was matched to 3-hydroxybenzaldehyde oxime.

fully matches the two. In this case the library again has no record for the correct compound.

When identification has been made by a computerized search, the analyst must obtain the speculative metabolite and confirm the identity based on chromatographic behavior as well as spectroscopy. As with EI GC/MS, compounds may decompose in the PB mass spectrometer source, occasionally giving spectra of decomposition products. Glucuronide and sulfate conjugates, for example, are prone to decomposition in PB LC/MS—the decomposition products are readily identified by computerized search as the corresponding aglycones.[63]

Quantitation in PB LC/MS

In addition to qualitative accuracy, reliable quantitation is an important consideration in biomonitoring. Quantitative performance is ultimately defined by accuracy and precision and method detection limits established by the analysis of actual samples. Instrument performance characteristics such as dynamic response, linear dynamic range, stability, instrument detection limits, and freedom from matrix effects, of course, influence quantitative performance. Because commercial PB LC/MS instruments have been available for only a few years, many of these performance characteristics are only now being investigated and reported.

Bellar and co-workers first observed that coeluting compounds enhance ion abundances in particle beam LC/MS.[17] They suggested two possible mechanisms for the signal enhancement: (1) the coeluting compound might reduce the vaporization of analytes from aerosol droplets and (2) the coeluting compound might reduce analyte losses in the momentum separator. There is considerable room for improving PB transport efficiency, which is generally less than 10% due to losses of volatile analytes, sedimentation of particles, nozzle and skimmer cone misalignment, and turbulence.[58] Bellar and co-workers referred to the signal enhancement phenomenon as a "carrier process" and associated it with PB nonlinearity. They concluded further that the carrier process had "negative implications" for the success of both conventional and isotope dilution quantitation with the new technique.[17]

In practice analytical chemists have taken advantage of the carrier effect to improve both the linearity and sensitivity of PB instruments. Continuous addition of malic acid,[13] phenylurea,[16] and phenoxyacetic acid[15] to the mobile phase improves the linearity of PB LC/MS in the determination of daminozide and phenylurea and chlorophenoxy acid herbicides, respectively. A structural analogue of the analyte may not be necessary as ammonium acetate appears to extend the linear range of many compounds. Kim and co-workers also observed spectral changes on addition of carriers[13] in isobutane positive CI PB LC/MS, but there is no evidence of this occurring in EI PB LC/MS.

Reliable quantitation is achievable in PB LC/MS when the instrument is calibrated properly and samples are analyzed within the calibrated range. First, PB LC/MS is linear for some compounds, but only over a limited range. Doerge and Miles[14] report a linear response for ethylenethiourea between 1.0 and about 100 ng without addition of a linearizing agent like ammonium acetate. Similarly, PB LC/MS has a linear response to siduron between 10 and 440 ng when using ammonium acetate as a carrier.[64] For many compounds, however, the instrument detection limits exceed ETU's and siduron's demonstrated linear dynamic range. Behymer and co-workers reported instrument detection limits (3:1 signal-to-noise ratio) ranging between 10 and 440 ng for a 16-compound mixture.[64] PB LC/MS nonlinearity therefore is the general rule, even with added carriers.

Determination of Styrene Metabolites by Isotope Dilution PB LC/MS

Styrene is excreted in urine in the form of three major metabolites, phenylglyoxylic (PGA), mandelic (MA), and hippuric (HA) acids. Two of these three metabolites, PGA, and MA, are determinants for occupational styrene exposure.[3] We evaluated PB LC/MS as an alternative method for determining styrene's metabolites in urine. Even with high concentrations of ammonium acetate in the mobile phase, PB LC/MS calibration curves for these metabolites are nonlinear (Fig. 11.10). Coeluting, stable-isotope labeled internal standards (the pentadeutero analogues) functioned as carriers at high concentrations (e.g., 0.5–5 µg),[62] but did not linearize the response of the native (unlabeled) analytes. Interestingly, we observed that the response of each isotope-labeled internal standard increased linearly with the amount of coeluting native compound. Thus, PB LC/MS is sensitive to a matrix effect, and any coeluting matrix constituents could conceivably affect quantitation of target compounds.

Fortunately, isotope dilution calibration curves are linear over a wide range and can compensate for the matrix effect. In Figure 11.10 isotope dilution calibration curves for styrene's urinary metabolites are shown for the range of 20–1500 ng/µL. The amount of labeled internal standard added was not critical between 50 and 500 ng/µL, and similar linear calibrations were observed. Urine samples from subjects exposed to styrene vapor in a laboratory chamber were also analyzed by isotope dilution PB LC/MS demonstrating the practical utility of this method. We concluded, unlike the earlier work of Bellar and co-workers,[17] that isotope dilution PB LC/MS is both practical and necessary for reliable quantitation.

CONCLUSIONS

Liquid chromatography/mass spectrometry techniques are potentially useful in human biomonitoring, particularly in the determination of polar metabolites, which serve as internal dose markers for many compounds. TSP LC/MS is well suited to the direct determination of metabolic conjugates including glucuronides, sulfates, and mercapturates. Strong anion-exchange columns, especially those based on styrene–divinylbenzene resins, eluted with acetate buffers are well suited to both separation and TSP mass spectrometry of these

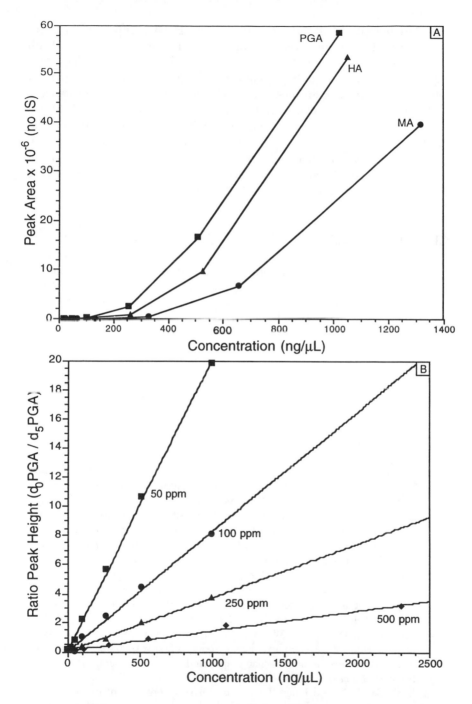

Figure 11.10 PB LC/MS calibration curves for PGA (■), MA (●), and HA (▲). Conventional (A) and isotope dilution calibration curves for PGA with increasing concentrations of internal standard (B) are shown. (Reproduced by permission of John Wiley & Sons Limited.[62])

metabolic conjugates. TSP coupled to tandem MS is the most powerful LC/MS instrument for identification and specific detection of polar metabolites. With the appropriate library, PB LC/MS has the capability to screen biological samples for an array of nontarget compounds and thus may find important applications in screening for metabolites of drugs and other toxic substances.

More research is needed to fully develop both PB and TSP LC/MS as routine quantitative techniques. While LC/MS is capable of direct determination of many involatile metabolite classes, GC/MS is clearly superior for quantitative analysis. Internal standards or better yet coeluting, stable isotope-labeled internal standards will be particularly useful in compensating for both the variability of TSP ionization, and the PB LC/MS matrix effect.

ACKNOWLEDGMENTS

We thank the staff of the California Department of Health Services Hazardous Materials Laboratory for their assistance and cooperation. This study was supported in part by the Superfund Program Project No. 04705 from the National Institute of Environmental Health Sciences of the National Institutes of Health.

REFERENCES

1. Hulka, B. S. and Wilcosky, T. Biological markers in epidemiological research. *Arch. Environ. Health* 43, 83, 1988.
2. Perera, F. P. and Weinstein, I. B. Molecular epidemiology and carcinogen-DNA adduct detection: New approaches to studies of human cancer causation. *J. Chron. Dis.* 35, 581, 1982.
3. American Conference of Governmental Industrial Hygienists. *Threshold Limit Values for Chemical Substances and Physical Agents and Biological Exposure Indices.* ACGIH, Cincinnati, OH, 1990–1991.
4. Cairns, T. and Siegmund, E. G. Review of the development of liquid chromatography/mass spectrometry. *Liquid Chromatography/Mass Spectrometry: Applications in Agricultural, Pharmaceutical, and Environmental Chemistry.* Brown, M. A., Ed. American Chemical Society, Washington, D.C., 1990, Ch. 1.
5. Chiu, K. S., Van Langenhove, A., and Tanaka, C. High-performance liquid chromatographic/mass spectrometric and high-performance liquid chromatographic/tandem mass spectrometric analysis of carbamate pesticides. *Biomed. Environ. Mass Spectrom.* 18, 200, 1989.
6. Voyksner, R. D., Pack, T. W., Haney, C. A., Freeman, H. S., and Hsu, W. N. Determination of the photodegradation products of Basic Yellow 2 by thermospray high-performance liquid chromatography and gas chromatography/mass spectrometry. *Biomed. Environ. Mass Spectrom.* 18, 1079, 1989.
7. Betowski, D. L. and Jones, T. L. The analysis of organophosphorus pesticide samples by high-performance liquid chromatography/mass spectrometry and high-performance liquid chromatography/mass spectrometry/mass spectrometry. *Environ. Sci. Technol.* 22, 1430, 1988.

8. Voyksner, R. D. and Haney, C. A. Optimization and application of thermospray high-performance liquid chromatography/mass spectrometry. *Anal. Chem.* 57, 991, 1985.

9. Wright, L. H., Jackson, M. D., and Lewis, R. G. Determination of aldicarb residues in water by combined high performance liquid chromatography/mass spectrometry. *Bull. Environ. Contam. Toxicol.* 28, 740, 1982.

10. Liu, C. H., Mattern, G. C., Xiaobing, Y., and Rosen, J. D. Determination of benomyl by high-performance liquid chromatography/mass spectrometry/selected ion monitoring. *J. Agric. Food Chem.* 38, 167, 1990.

11. Voyksner, R. D., Bursey, J. T., and Pellizzari, E. D. Analysis of selected pesticides by high-performance liquid chromatography-mass spectrometry. *J. Chromatogr.* 312, 221, 1984.

12. Parker, C. E., Haney, C. A., and Hass, J. R. High-performance liquid chromatography-negative chemical ionization mass spectrometry of organophosphorus pesticides. *J. Chromatgr.* 237, 233, 1982.

13. Kim, I. S., Sasinos, F. I., Stephens, R. D., and Brown, M. A. Analysis of daminozide in apple juice by anion-exchange chromatography-particle beam mass spectrometry. *J. Agric. Food Chem.* 38, 1223, 1990.

14. Doerge, D. R. and Miles, C. J. Determination of ethylenethiourea in crops using particle beam liquid chromatography/mass spectrometry. *Anal. Chem.* 63, 1999, 1991.

15. Mattina, M. J. I. Determination of chlorophenoxy acids using high-performance liquid chromatography-particle beam mass spectrometry. *J. Chromatogr.* 542, 385, 1991.

16. Mattina, M. J. I. Carrier effect in the analysis of phenylurea herbicides using high-performance liquid chromatography-particle beam-mass spectrometry. *J. Chromatogr.* 549, 237, 1991.

17. Bellar, T. A., Behymer, T. D., and Budde, W. L. Investigation of enhanced ion abundances from a carrier process in high-performance liquid chromatography particle beam mass spectrometry. *J. Am. Soc. Mass Spectrom.* 1, 92, 1990.

18. Dulik, D. M., Colvin, O. M., and Fenselau, C. Characterization of glutathione conjugates of chlorambucil by fast atom bombardment and thermospray liquid chromatography/mass spectrometry. *Biomed. Environ. Mass Spectr.* 19, 248, 1990.

19. Voyksner, R. D., Hagler, W. M., and Swanson, S. P. Analysis of some metabolites of T-2 toxin, diacetoxyscirpenol and deoxynivalenol by thermospray high-performance liquid chromatography-mass spectrometry. *J. Chromatogr.* 394, 183, 1987.

20. Kostiaianen, R. and Kuronen, P. Use of 1-[p-(2,3-dihydroxypropoxy)phenyl]-1-alkanones as retention index standards in the identification of trichothecenes by liquid chromatography thermospray and dynamic fast atom bombardment mass spectrometry. *J. Chromatogr.* 543, 39, 1991.

21. McManus, K. T., deBethizy, J. D., Garteiz, D. A., Kyerematen, G. A., and Vesell, E. S. A new quantitative thermospray LC-MS method for nicotine and its metabolites in biological fluids. *J. Chromatogr. Sci.* 28, 510, 1990.

22. Blaszkewicz, M., Baumhoer, G., Neidhart, B., Ohlendorf, R., and Linscheid, M. Identification of trimethyllead in urine by high-performance liquid chromatography with column switching and chemical reaction detection and by liquid chromatography-mass spectrometry. *J. Chromatogr.* 439, 109, 1988.

23. Spink, D. C., Aldous, K. M., and Kaminsky, L. S. Analysis of oxidative warfarin metabolites by thermospray high-performance liquid chromatography/mass spectrometry. *Anal. Biochem.* 177, 307, 1989.

24. Bieri, R. H. and Greaves, J. Characterization of benzo[*a*]pyrene metabolites by high performance liquid chromatography-mass spectrometry with a direct liquid introduction interface and using negative chemical ionization. *Biomed. Environ. Mass Spectrom.* 14, 555, 1987.

25. Dragna, S., Aubert, C., and Cano, J. P. Direct liquid inlet liquid chromatographic/mass spectrometric identification and high-performance liquid chromatographic analysis of a benzodiazepine glucuronide. *Biomed. Environ. Mass Spectrom.* 18, 359, 1989.

26. Yinon, J. and Hwang, D. G. Metabolic studies of explosives. II. High-performance liquid chromatography-mass spectrometry of metabolites of 2,4,6-trinitrotoluene. *J. Chromatogr.* 339, 127, 1985.

27. Lant, M. S., Oxford, J., and Martin, C. E. Automated sample preparation on-line with thermospray high-performance liquid chromatography-mass spectrometry for the determination of drugs in plasma. *J. Chromatogr.* 394, 223, 1987.

28. Bowers, L. D. High-performance liquid chromatography/mass spectrometry: State-of-the-art for the drug analysis laboratory. *Clin. Chem.* 35, 1282, 1989.

29. Brown, F. R. and Draper, W. M. Separation of phenols and their glucuronide and sulfate conjugates by anion exchange liquid chromatography. *J. Chromatogr.* 479, 441, 1989.

30. Johnston, J. J., Draper, W. M., and Stephens, R. D. LC-MS compatible HPLC separation for xenobiotics and their phase I and phase II metabolites: Simultaneous anion exchange and reversed phase chromatography. *J. Chromatogr. Sci.* 29, 511, 1991.

31. Blakely, C. R., Carmody, J. J., and Vestal, M. L. A new ionization technique for mass spectrometry of complex molecules. *J. Am. Chem. Soc.* 102, 5931, 1980.

32. Fenselau, C. and Yellet, L. In *Xenobiotic Conjugation Chemistry.* Paulson, G., Cladwell, J., Hutson, D., and Menn, J., Eds. ACS Symposium Series No. 299. American Chemical Society, Washington, D.C., 1986, pp. 159–176.

33. Liberato, D. J., Fenselau, C. C., Vestal, M. L., and Yergey, A. L. Characterization of glucuronides with a thermospray liquid chromatography/mass spectrometry interface. *Anal. Chem.* 55, 1741, 1984.

34. Watson, D., Taylor, G. W., and Murray, S. Thermospray liquid chromatography negative ion mass spectrometry of steroid sulfate conjugates. *Biomed. Environ. Mass Spectrom.* 12, 610, 1985.

35. Personal communication from A. L. Yergey cited in Garteiz, D. A. and Vestal, M. L. Thermospray LC/MS interface: Principles and applications. *LC Mag.* 3, 1985.

36. Yergey, A. L., Liberato, D. J., and Millington, D. S. Thermospray liquid chromatography/mass spectrometry for the analysis of l-carnitine and its short-chain acyl derivatives. *Anal. Biochem.* 139, 278, 1984.

37. Barcelo, D., Durand, G., Vreeken, R. J., de Jong, G. J., and Brinkman, U. A. T. Nonpolar solvents for normal-phase liquid chromatography and postcolumn extraction in thermospray liquid chromatography/mass spectrometry. *Anal. Chem.* 62, 1696, 1990.

38. Parker, C. E., Smith, R. W., Gaskell, S. J., and Bursey, M. M. Dependence of ion formation upon the ionic additive in thermospray liquid chromatography/negative ion mass spectrometry. *Anal. Chem.* 58, 1661, 1986.

39. Alexander, A. J. and Kebarle, P. Thermospray mass spectrometry: Use of gas-phase ion/molecule reactions to explain features of thermospray spectra. *Anal. Chem.* 58, 471, 1986.
40. Smith, R. W. and Parker, C. E. Eluent pH and thermospray mass spectra: Does the charge of the ion in solution influence the mass spectrum? *J. Chromatogr.* 394, 261, 1987.
41. Blakely, C. R. and Vestal, M. L. Thermospray interface for liquid chromatography/mass spectrometry. *Anal. Chem.* 55, 750, 1983.
42. Voyksner, R. D. and Yinon, J. Trace analysis of explosives by thermospray high-performance liquid chromatography-mass spectrometry. *J. Chromatogr.* 354, 393, 1986.
43. Wyllie, J. E., Gabica, J., Benson, W. W., and Yoder, J. *Pestic. Monit. J.* 9, 150, 1975.
44. Shafik, M. T., Sullivan, H. C., and Enos, H. F. A method for determination of low levels of exposure to 2,4-D and 2,4,5-T. *Int. J. Environ. Anal. Chem.* 1, 23, 1971.
45. Edgerton, T. R. and Moseman, R. F. Determination of pentachlorophenol in urine: The importance of hydrolysis. *J. Agric. Food Chem.* 27, 197, 1979.
46. Draper, W. M. A multiresidue procedure for the determination and confirmation of acidic herbicide residues in human urine. *J. Agric. Food Chem.* 30, 227, 1981.
47. Jones, T. L., Betowski, L. D., and Yinon, J. Analysis of chlorinated herbicides by high-performance liquid chromatography/mass spectrometry. *Liquid Chromatography/Mass Spectrometry: Applications in Agricultural, Pharmaceutical, and Environmental Chemistry.* Brown, M. A., Ed. American Chemical Society, Washington, D.C., 1990, Ch. 5.
48. Wright, L. H., Edgerton, T. R., Arbes, S. J., and Lores, E. M. The determination of underivatized chlorophenols in human urine by combined high performance liquid chromatography mass spectrometry. *Biomed. Mass Spectrom.* 8, 475, 1981.
49. Blaszkewicz, M., Baumhoer, G., Neidhart, B., Ohlendorf, R., and Linscheid, M. Identification of trimethyllead in urine by high performance liquid chromatography with column switching and chemical reaction detection and by liquid chromatography-mass spectrometry. *J. Chromatogr.* 439, 109, 1988.
50. Draper, W. M., Brown, F. R., Bethem, R., and Miille, M. J. Thermospray mass spectrometry and tandem mass spectrometry of polar, urinary metabolites and metabolic conjugates. *Biomed. Environ. Mass Spectrom.* 18, 767, 1989.
51. Draper, W. M., Brown, F. R., Bethem, R., and Miille, M. J. Anion exchange thermospray tandem mass spectrometry of polar urinary metabolites and metabolic conjugates. *Liquid Chromatography/Mass Spectrometry: Applications in Agricultural, Pharmaceutical, and Environmental Chemistry.* Brown, M. A., Ed. American Chemical Society, Washington, D.C., 1990, Ch. 17.
52. Lowry, L. K., Tolos, W. P., Boeniger, M. F., Nony, C. R., and Bowman, M. C. Chemical monitoring of urine from workers potentially exposed to benzidine-derived azo dyes. *Toxicol. Lett.* 7, 29, 1980.
53. Dewan, A., Jani, J. P., Patel, J. S., Gandhi, D. N., Variya, M. R., and Ghodasara, N. B. Benzidine and its acetylated metabolites in the urine of workers exposed to Direct Black 38. *Arch. Environ. Health* 43, 269, 1988.
54. Straub, K. M., Rudewicz, P., and Garvie, C. Metabolic mapping of drugs: Rapid screening techniques for xenobiotic metabolites with m.s./m.s. techniques. *Xenobiotica* 17, 413, 1987.

55. Pearson, P. G., Howald, W. N., and Nelson, S. D. Screening strategy for the detection of derivatized glutathione conjugates by tandem mass spectrometry. *Anal. Chem.* 62, 1827, 1990.

56. Straub, K. M. *Mass Spectrometry in Biomedical Research.* Gaskell, S. J., Ed. John Wiley, New York, 1986, p. 115.

57. Willoughby, R. C. and Browner, R. F. Monodisperse aerosol generation interface for combining liquid chromatography with mass spectrometry. *Anal. Chem.* 56, 2626, 1984.

58. Winkler, P. C., Perkins, D. D., Williams, W. K., and Browner, R. F. Performance of an improved monodisperse aerosol generation interface for LC/MS. *Anal. Chem.* 60, 489, 1988.

59. McLafferty, F. W. *Interpretation of Mass Spectra.* Turro, N. J., Ed., Organic Chemistry Series. University Science Books, Mill Valley, CA, 1980, p. 232.

60. Shafik, M. T. and Enos, H. F. Determination of metabolic and hydrolytic products of organophosphorus pesticide chemicals in human blood and urine. *J. Agric. Food Chem.* 17, 1186, 1969.

61. Draper, W. M., Wijekoon, D., and Stephens, R. D. Determination of malathion urinary metabolites by isotope dilution ion trap GC/MS. *J. Agric. Food Chem.* 39, 1796, 1991 (and references cited therein).

62. Brown, F. R. and Draper, W. M. The matrix effect in particle beam liquid chromatography/mass spectrometry and reliable quantification by isotope dilution. *Biol. Mass Spectrom.* 20, 515, 1991.

63. Brown, F. R. and Draper, W. M. Particle beam liquid chromatography/mass spectrometry of phenols and their sulfate and glucuronide conjugates. *Liquid Chromatography/Mass Spectrometry: Applications in Agricultural, Pharmaceutical, and Environmental Chemistry.* Brown, M. A., Ed. American Chemical Society, Washington, D.C., 1990, Ch. 15.

64. Behymer, T. D., Bellar, T. A., and Budde, W. L. Liquid chromatography/particle beam/mass spectrometry of polar compounds of environmental interest. *Anal. Chem.* 62, 1686, 1990.

CHAPTER 12

Measurement of Reductive Dechlorination of Pentachlorophenol by *Actinomyces viscosus* Strain *dechlorini* by GC/MS Techniques

Frank O. Bryant and Horace G. Cutler

INTRODUCTION

Actinomyces viscosus strain *dechlorini* was isolated from a commercial plant growth stimulator derived from fermented dairy cow manure. The bacterium was purified by serial dilution and identified by systematic tests performed in accordance with *Bergey's Manual of Systematic Bacteriology*. The bacterium biotransformed pentachlorophenol (PCP) under anaerobic conditions to 2,3,5-trichlorophenol, 2,5-dichlorophenol, 2-chlorophenol, and phenol. Biotransformation of PCP by the bacterium required yeast extract and was also enhanced by DL-lactic acid. The rate of cell replication was increased as PCP was biotransformed, indicating that PCP may serve as an energy source for the bacterium. The bacterium was tested for its capacity to bioremediate site samples from a wood treatment facility that contained PCP and other chlorinated/aromatic compounds. In laboratory studies, the samples inoculated with the bacterium removed PCP and other contaminates at faster rates than non-inoculated control samples.

Bioremediation of hazardous chemicals is rapidly developing as a desirable method of cleaning up hazardous waste sites at lower cost compared to conventional remediation methods.[1] Also, conventional remediation techniques, such as excavation, filtering, or immobilization, are not necessarily designed to transform contaminants into innocuous products. Instead, such techniques typically contain and/or concentrate only the hazardous compound(s).[2] By contrast, bioremediation offers the capacity to mineralize contaminating compounds to fermentative endproducts, CO_2 and/or CH_4.[3,4] The USEPA and others have invested heavily in research to develop bioremediation schemes for numerous toxic and mutagenic chemicals that exist in the environment, including chlorinated aromatic compounds.[5-12] Homocyclic, aryl chlorides are of low solubility and are recalcitrant to natural biodegradation due to chlorination of the aromatic ring.[13] Consequently, emphasis has been placed on bioremediation schemes that dechlorinate the aromatic ring, increasing solubility and, thereby, exposure to microbiological degradation.

Pentachlorophenol (PCP) is a toxic and mutagenic chlorinated aromatic compound that is still routinely used as a biocide/fungicide for wood preservation in the U.S.[14-16] Outside the U.S., PCP use is even more widespread.[17] Consequently, the USEPA has classified PCP as a priority pollutant.[18] Numerous bacteria have been isolated in pure culture that can biodegrade PCP and other chlorinated aromatic compounds under aerobic conditions.[19-28] However, attempts to biodegrade PCP in the presence of O_2 can result in either no PCP removal[9] or generation of pentachloroanisole as the main product.[13,29] Although pentachloroanisole is less toxic than PCP, it is a more stable compound and consequently less accessible for further biodegradation.

By comparison, PCP biodegradation under anaerobic conditions proceeds predominantly by the mechanism of reductive dechlorination as the initial step.[13] Reductive dechlorination involves a two electron transfer in which chlorine is removed as chloride (Cl^-) from homocyclic aryl chlorides and replaced by a proton (H^+) from water.[30] Reductive dechlorination of chlorinated aromatic compounds by microbial consortia has been well studied in sludges and sediments under methanogenic-type, anaerobic conditions.[5,8,12,31-37] A period of adaptation to chlorophenols is often required, however, to establish effective rates of mineralization in such undefined microbial consortia. An excellent summary of pathways of reductive dechlorination of chlorophenols by nonadapted and adapted microbial consortia was recently published.[36]

One anaerobic bacterium isolated in pure culture, *Desulfomonile tiedjei* DCB-1, removes chlorine and derives energy and carbon from 4-chlorobenzoic acid.[38] Reductive dechlorination of 4-chlorobenzoic acid proceeds by reduction of the aromatic ring and the addition of a proton derived from water.[40,41] DCB-1 also dechlorinates PCP at low concentrations (<3 μg/mL) but only as a detoxifying mechanism since neither energy nor carbon is derived from PCP.[9] *D. tiedjei*, taxonomically, is a sulfate reducer, and the reduction of sulfate is preferred over dechlorination as a means of removing reducing equivalents.[39] However, sulfate reduction and dechlorination apparently share components of an electron transport chain in DCB-1. Consequently, reductive dechlorination may constitute a novel form of anaerobic respiration.

ISOLATION AND TAXONOMIC CHARACTERISTICS OF *ACTINOMYCES VISCOSUS* STRAIN *DECHLORINI*

Recently, a bacterium tentatively designated *Actinomyces viscosus* strain *dechlorini* was isolated in our laboratory from a commercial plant growth

Table 12.1 Minimal Medium for Isolation of *Actinomyces viscosus* Strain *dechlorini* under Anaerobic Conditions for Reductive Dechlorination of Pentachlorophenol[a]

Ingredients	Quantities/L
KH$_2$PO$_4$	1.5 g
K$_2$PO$_4$	1.5 g
NH$_4$Cl	0.5 g
MgCl$_2$ · 6H$_2$O	0.18 g
Yeast extract (Difco)	2.0 g
Glucose[b]	8.0 g
Reducing solutions[c]	40.0 mL

[a] The medium was prepared under anaerobic conditions using the Hungate technique.[44] The ingredients of the medium were combined and ~30 mL distributed into 50-mL serum-stoppered bottles. The serum-stoppered bottles were vented with disposable 23-gauge syringe needles and autoclaved for 50 min at 15 psi. The serum-stoppered bottles were removed immediately after the autoclave cycles was complete and flushed with N$_2$ gas (>99%) for at least 10 min. The medium was used immediately upon cooling to room temperature. Inoculation was performed anaerobically with a syringe from a growing culture of the bacterium. PCP (10 µg/mL) was added with a syringe from a stock solution of 3000 µg/mL PCP in 50% aqueous methanol. The bacterium was unable to utilize methanol as a growth substrate.

[b] Deleted from medium after bacterium was isolated.

[c] Reducing solution contained 200 mL of 0.2 N NaOH plus 2.5 g Na$_2$S · 9H$_2$O.

stimulator derived from fermented dairy cow manure.[42] The plant growth stimulator was a mixed consortium of various microorganisms and was previously suggested to have the capacity to consume oil, sewage, and PCBs. The bacterium was purified by serial dilution (1% transfers through 10 serum-stoppered bottles repeated twice) using the medium shown in Table 12.1. The bacterium was subjected to systematic tests, including procedures for testing special characteristics, performed in accordance with *Bergey's Manual of Systematic Bacteriology*[42] and *Manual of Methods for General Bacteriology*.[43]

Specifically, the bacterium grew under anaerobic and aerobic conditions indicating that it was a facultative anaerobe. Microcolonies grew aerobically within 24 hr on potato dextrose agar (PDA) at 37°C displaying a dense core and filamentous (spidery) appearance at the periphery. Liquid cultures initially displayed filamentous cells of various lengths and branching. As the culture grew, the filaments eventually fragmented into coccoid cells and finally formed mucoid aggregates that appeared as a flocculent mass in the medium. Consequently, the bacterium displayed pleomorphic growth. All tests for spores were negative, although coccoid cells were refractile in appearance. The bacterium was indole negative, urease negative, gas production negative, and catalase positive. The principal fermentation product was lactic acid, but acetic, succinic, and formic acid were also produced; propionic acid was not produced. All cell types were nonmotile. These findings are summarized in Table 12.2 and are

Table 12.2 Characteristic of *Actinomyces viscosus* strain *dechlorini*

Facultative anaerobe	
Pleomorphic	Filaments (early log phase), diphtheroid or chains of coccoids (mid to late log phase), coccoids and aggregates (late log and stationary phase); all cell types are nonmotile.
Microcolonies	Dense core with filamentous (spidery) periphery, no pigments
Liquid culture	Log phase cultures display heavy growth with a clear zone between liquid surface and ~1 cm depth into medium; late log and stationary phase culture displays mucoid or papery cell aggregates.
Optimal growth temperature range	20–37°C
Gram stain and acid fast responses	Positive
Spore tests	Negative
Indole, urease, and gas production	Negative
Catalase activity	Positive
Principal fermentation product from glucose	Lactate
Minor products	Acetate, succinate, and formate but *not* propionate
Principal fatty acids	C_{15} and C_{17}

consistent with our tentative taxonomic classification of the bacterium within the family Actinomycetaceas and the genus *Actinomyces*.[45] Based on a positive castalase reaction, tentative assignment of the bacterium to the species *viscosus* was also made.[46]

BIOTRANSFORMATION OF PCP BY *A. VISCOSUS* STRAIN *DECHLORINI*

Minimal Medium for PCP Biotransformation by *A. viscosus* Strain *dechlorini*

To maximize PCP biotransformation by *A. viscosus* strain *dechlorini*, the least complex medium was determined in which to culture the bacterium to facilitate the utilization of PCP as a carbon and/or energy source. As shown in Figure 12.1, PCP (13 µg/mL) was biotransformed by a culture of *A. viscosus* strain *dechlorini* under anaerobic conditions at 37°C during 24 hr growth on a medium of basic mineral salts (Table 12.1) and yeast extract (0.1%) (glucose was excluded from the medium). Upon a second addition of PCP (52 µg/mL), PCP was removed to 5.5 µg/mL. Cell growth increased from 2×10^6 to 2×10^8 cell/mL as PCP was removed. No lag phase was associated with PCP removal. Upon transfer of the bacterium to fresh medium containing only basic salts and reducing solution (Fig. 12.2), PCP (12 µg/mL) was biotransformed at a slower rate requiring 13 days for removal at 37°C. Growth of the bacterium increased from 10^6 to 10^9 cell/mL as PCP was removed. Although the rate of PCP biodegradation was lower than in Figure 12.1, no lag phase was apparent and the final concentration of cells was higher. Both cultures actively biotransformed

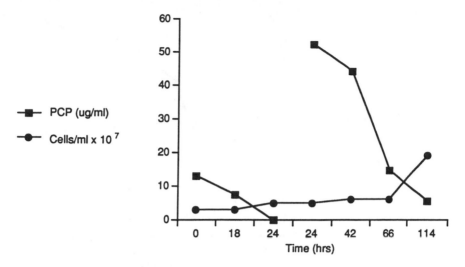

Figure 12.1   Progress of biotransformation of PCP and cell growth of *A. viscosus* strain *dechlorini*
on a medium of basic mineral salts and yeast extract (0.1%). The inoculum (0.1 mL)
was from the final bottle of the serial dilution used to obtain the bacterium in pure
culture. PCP (99%) was obtained from Aldrich Chemical Co. A stock solution of PCP
(3000 µg/mL) in 50% aqueous methanol was used to supply PCP to cell cultures and
controls. Methanol was not utilized as a growth substrate by the bacterium. PCP was
identified and quantified by HPLC analysis and cochromatography with a PCP
standard in 50% aqueous methanol. The HPLC system consisted of an LDC
ConstaMetric 4100 solvent delivery system, an LDC SpectroMetric 5000 photodiode
array monitor selected at 290 nm, a Gateway 2000 computer equipped with LDC
software package, and an Epson LX-810 recorder. The column was a nucleosil C$_{18}$ 5U
reverse-phase (250 × 4.6 mm). The mobile phase was ethanol:water:glacial acetic acid
(80:19:1). Cell growth was monitored by direct cell count using an Improved Neubauer
Ultra Plane counting chamber (1/400 mm^2 × 1/10 mm) and an Olympus BH2 phase
contrast microscope.

PCP and accumulated 2,4,5-trichlorophenol (TCP) as determined by comparison to standards. Tetrachlorophenols (TeCP) were not detected in sufficient
amounts to be identified. Determination of lesser chlorinated intermediates was
not performed for cultures depicted in Figures 12.1 and 12.2.

The observed difference in rate of PCP removal between Figures 12.1 and
12.2 is likely the result of added yeast extract since this is the only difference in
the media. The culture described in Figure 12.2 was limited to the yeast extract
obtained in the inoculum from the culture in Figure 12.1. A subsequent attempt
to obtain growth and PCP biotransformation without yeast extract in the medium using the culture of Figure 12.2 as inoculum displayed little cell growth
or PCP biotransformation (not shown). Several concentrations of glucose in the
minimal medium promoted excellent growth of the bacterium, but PCP was
either not removed or removed slowly when added to glucose-containing media
(not shown). Autoclaved control cultures were maintained for each active culture. Controls did not remove PCP, accumulate intermediates, or demonstrate
cell replication.

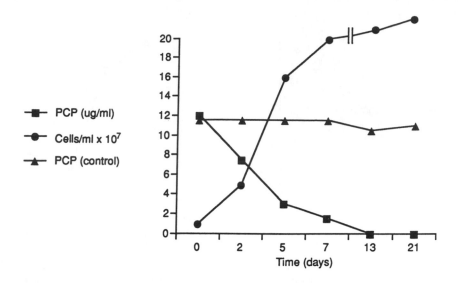

Figure 12.2   Progress of biotransformation of PCP and cell growth on a medium of basic mineral
salts. The inoculum (0.2 mL) was from the active culture of Figure 12.1. All other
conditions are as in Figure 12.1.

Enhanced PCP Biotransformation by *A. viscosus* Strain *dechlorini* with DL-Lactic Acid Supplementation

To enhance biotransformation of PCP at concentrations above 100 µg/mL by *A. viscosus* strain *dechlorini,* the basic mineral salts medium containing yeast extract (0.1%) was supplemented with various metal salts and volatile fatty acids. As shown in Figure 12.3, supplementation of the medium with DL-lactic acid biotransformed PCP at an initial concentration of 126 µg/mL at 22°C. As PCP was biotransformed between days 0 to 5, cell replication increased logarithmically to 2.5×10^8 cells/mL. A subsequent addition of DL-lactic acid (0.17%) on day 9 resulted in a second phase of logarithmic cell growth to $>10^9$ cells/mL. During the time course, chlorinated intermediates including 2,3,5-TCP, 2,5-dichlorophenol (DCP), and 2-chlorophenol (CP) were produced. Biotransformation of PCP at concentrations above 100 µg/mL by the bacterium without DL-lactic acid supplementation either did not occur or was sufficiently slow that intermediates were not produced.

Figure 12.4a shows the progress of cell replication at 22°C of *A. viscosus* strain *dechlorini* grown in the presence of PCP compared to a culture of the bacterium grown on the same medium but without PCP. The cultures were inoculated and monitored coincidentally. A higher rate of cell replication was observed in the culture containing PCP until day 13 when both cultures reached a concentration of 6×10^8 cells/mL. Upon a second addition of DL-lactic acid on day 13, a higher rate of cell replication was again observed in the culture containing PCP until day 19 when both cultures reached concentration of 1.2

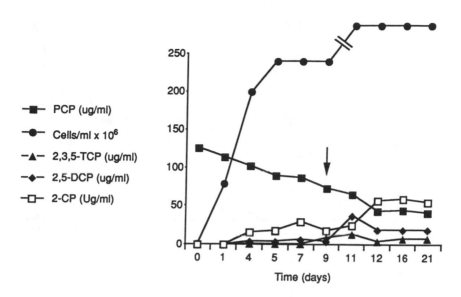

Figure 12.3   Progress of biotransformation of PCP and product formation in a medium of basic
mineral salts, 0.1% yeast extract, and 0.05% DL-lactic acid (85%) with a subsequent
addition of DL-lactic acid (0.17%) on day 9, designated by arrow. The inoculum (0.3
mL) was from the active culture of Figure 12.2. Cell concentrations on days 11,12,16,
and 21 were ~10⁹ cells/mL. PCP and 2,3,5-TCP were identified and quantified as
described in Figure 12.1. 2,5-DCP and 2-CP were identified and quantified as
described in Figure 12.1 except that the mobile phase was acetonitrile:water:glacial
acetic acid (50:48:2) and the photodiode array monitor was selected at 280 nm. A
chlorinated phenol kit from Ultra Scientific containing all chlorophenol congeners
was obtained for cochromatography to identify and quantify intermediates from
reductive dechlorination of PCP. Cell growth was as determined in Figure 12.1.

$\times 10^9$ cell/mL. Both cultures demonstrated cell aggregation beyond day 19, making accurate cell counts difficult.

Figure 12.4b shows the progress of biotransformation of PCP and product formation by the *A. viscosus* strain *dechlorini* of Figure 12.4a. Similar to Figure 12.4a, the addition of DL-lactic acid on day 13 was coincident with an increase in the rate of biotransformation of PCP (19–1 µg/mL). A subsequent addition of PCP on day 15 was biotransformed at a similar rate by day 19 (123–71 µg/mL) when further removal of PCP ceased. DL-Lactic acid (0.16%) was again added on day 21, but an increase in the rate PCP biotransformation was not observed until between day 28 and 41 (74–8 µg/mL). Since the final addition of DL-lactic acid was not coincident with an increase in the rate of PCP biotransformation and since a dependence on yeast extract was previously noted, additions of yeast extract were made on days 39 (0.1%) and 43 (0.1%). Coincident with yeast extract addition was an increase in the biotransformation of 2,3,5-TCP between days 39 and 41 (75–10 µg/mL) and, subsequently, 2-CP between days 41 and 43 (64–1 µg/mL). Phenol was observed by day 39 coincident with accumulation of 2,3,5-TCP and 2-CP. An autoclaved culture was monitored as a control. PCP was not biotransformed nor were chlorinated intermediates formed in the control.

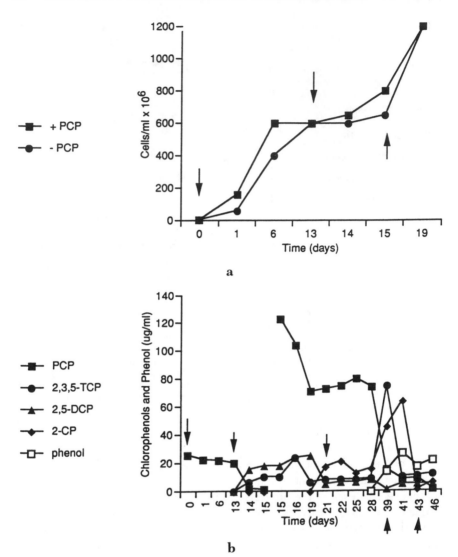

Figure 12.4   (a) Effect of PCP on *A. viscosus* sp. strain *dechlorini* cell replication. Progress of cell replication was monitored as described in Figure 12.1. The bacterium was grown on minimal media containing yeast extract (0.1%) (but not glucose) with additions of DL-lactic acid on day 0 (0.1%) and on day 13 (0.16%), indicated by arrows. PCP additions were on day 0 (26 μg/mL) and on day 15 (123 μ/mL), indicated by arrows. A separate culture was monitored for cell replication to which PCP was never added. The inoculum was 0.1 mL of the culture actively biotransforming PCP described in Figure 12.3. (b) Progress of biotransformation of PCP and product formation by the *A. viscosus* sp. strain *dechlorini* culture described in (a). Further additions of DL-lactic acid on day 21 (0.16%) and of yeast extract on day 39 (0.1%) and on day 43 (0.1%) are indicated by arrows. PCP and products were identified and quantified as described in Figures 12.1 and 12.3. An autoclaved culture was maintained as a control. PCP was not removed nor were chlorinated intermediates produced in the control.

A separate culture similar to that of Figure 12.4b was analyzed at specified intervals by GC/MS. As shown in Figure 12.5, on day 1 only PCP was identified. By day 19, PCP and phenol were identified by comparison to the NSF data base. Other intermediates were apparent from the GC chromatogram. By day 40, PCP concentration was lower and the intermediates observed on day 19 were removed. Phenol was also identified, but at a lower relative concentration.

BIOTRANSFORMATION OF PCP-CONTAINING AQUEOUS SAMPLES FROM MEREDITH WOOD TREATMENT FACILITY

Laboratory Scale Biotransformation of PCP-Containing Aqueous Samples

On-site samples were obtained (March 5, 1992) from the W.C. Meredith Co. wood treatment facility (Atlanta, GA). The samples included unflocked process water and flocked process water in which PCP was the major aromatic contaminant. Phenanthrene, fluoranthrene, n-docosane, n-tetracosane, pyrene, and hexanoic acid were also present in the process water. Also, a sample of groundwater from the site contained PCP and numerous other halogenated and aromatic compounds including nitrosopyrrolidine, pentchlorobenzene, nitrophenol, dibenzofuran, fluroanthene, fluorene, nitrobenzene, nitroaniline, nitrosopiperidine, toxaphene, PCBs, 2,4-D, and 2,4,5-T as major contaminants as determined by an independent laboratory. Aliquots (30 mL) of each sample were transferred to separate 50-mL serum-stoppered bottles, autoclaved for 20 min at 15 psi, and degassed with N_2 to obtain anoxic conditions. All samples were made to 0.5% DL-lactic acid. One bottle from each sample source was inoculated with A. viscosus strain dechlorini. A noninoculated control was maintained from each sample source.

Table 12.3 shows the three aqueous sample types monitored for PCP transformation at specific intervals. Both control and active samples from each source transformed PCP. However, the rate of PCP biotransformation in each sample type was greater in the aliquot that was inoculated with the bacterium than in the noninoculated control. In the case of the unflocked and flocked process water, PCP was completely removed by day 22 while controls contained residual PCP. Also, Figure 12.6a and b shows the HPLC chromatogram of the active samples of the unflocked and flocked on days 0 and 22. The chromatogram on day 0 indicated the presence of PCP and numerous other minor components. By day 22, not only was PCP absent but the minor components were absent as well.

Similarly for the groundwater sample, Figure 12.7 shows the HPLC chromatograms. The active sample indicated the presence of PCP and numerous major and minor components on day 0. On days 13 and 22, PCP and most of the major and minor components were progressively removed.

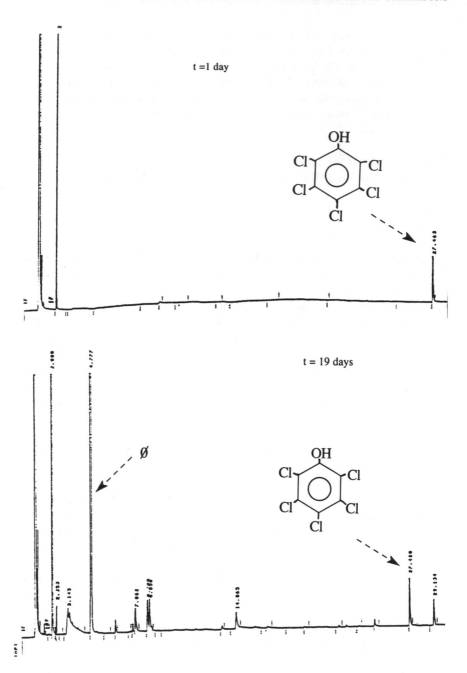

Figure 12.5

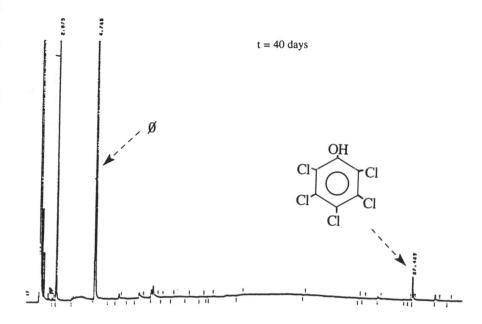

Figure 12.5   GC/MS analysis of the progress of dechlorination by a culture of *A. viscosus* sp. strain *dechlorini* similar to that described in Figure 12.4b. Culture samples (3 mL) were removed on days 1, 19, and 40 and combined with diethyl ether (3 mL) to extract chlorinated phenols and phenol. The ether fraction was concentrated and analyzed with an Extrel Model C50/400 quadrupole mass spectrometer interfaced with a Perkin-Elmer Sigma 300 gas–liquid chromatograph equipped with a cold on-column injector. Operating conditions were essentially as previously described,[47] except that the silica capillary column was 45 m in length and that the carrier gas pressure was 6 psi. Compounds were identified by matching their mass spectrum with those of the NIST/EPA mass spectral data base.

Table 12.3   **PCP Biotransformation in Aqueous Samples from Meredith Wood Treatment Facility Inoculated with *A. viscosus dechlorini*[a]**

	PCP—control/active (μg/mL)		
Time (days)	MW1[b]	MW2[b]	WCM4[b]
0	592.3	240.2	136.8
8	185.4/79.2	35.6/17.9	37.7/20
13	160.1/35.3	30.9/0	6.1/2.9
22	17.4/0	39.3/0	0/0

[a] All samples were collected on March 5, 1992. The samples were placed in 50-mL serum-stoppered bottles, autoclave 20 min, and degassed with N_2. To each sample was added 0.5% DL-lactic acid (85%). Control samples were analyzed without further treatment. Active samples were inoculated with 1 mL of *A. viscosus dechlorini* actively dechlorinating PCP. The samples were maintained on the benchtop at ~22°C. Samples were taken at specified intervals, filtered through a 0.2-μm filter (Millipore), and analyzed by HPLC as in Figure 12.1.

[b] MW1: unflocked process water (pH 5.5); MW2: flocked process water (pH 6); WCM4: groundwater (pH 6).

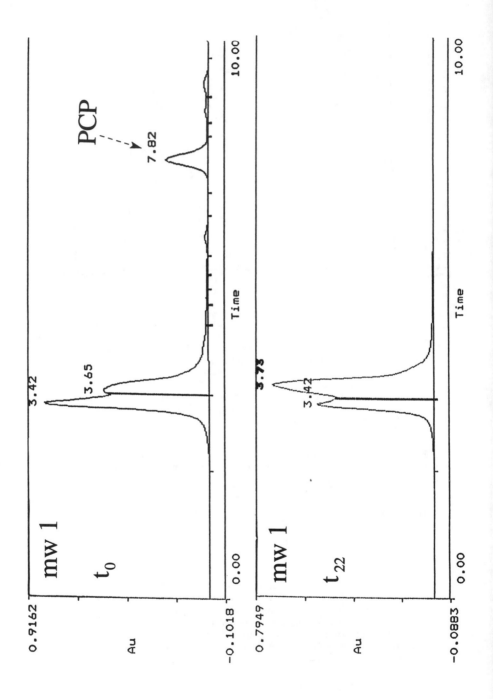

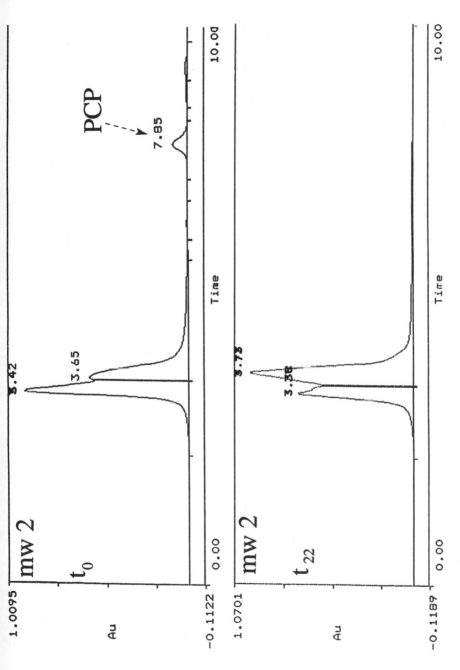

Figure 12.6 (a) HPLC chromatograms of unflocked process water from Meredith wood treatment facility inoculated with *A. viscosus* sp. strain *dechlorini* on days 0 and 22 as described in Table 12.3. (b) HPLC chromatogram of flocked process water as described in (a).

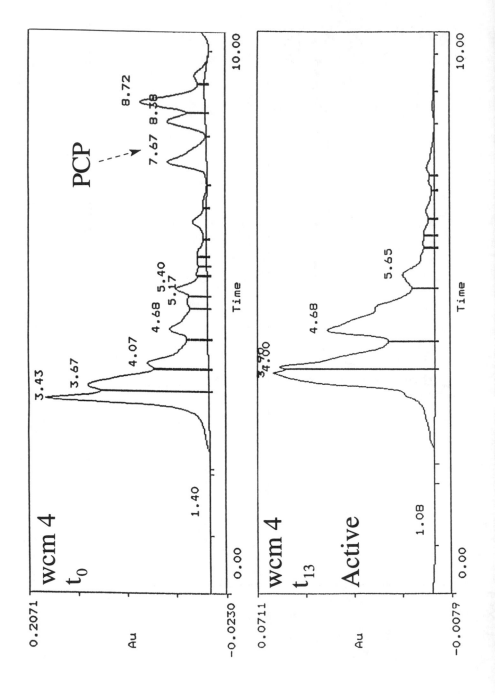

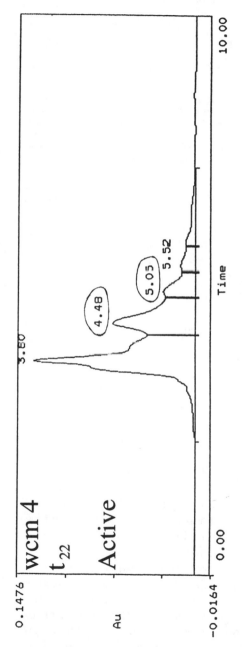

Figure 12.7 HPLC chromatograms of groundwater from Meredith wood treatment facility inoculated with *A. viscosus* sp. strain *dechlorini* on days 0, 13, and 22 as described in Table 12.3.

Table 12.4   PCP Biotransformation in Flocked Process Water on Site at Meredith
             Wood Treatment Facility[a]

Time	Concentration—control/active (µg/mL)					
(days)	PCP	2,3,4,6-TeCP	2,4,6-TCP	2,4-DCP	2-CP	Phenol
0	380	20	5	0/0	5	5
1	340/280	5/7	5/7	0/0	5/7	5/7
9	170/150	20/20	10/10	0/0	10/10	10/10
14	118/140	0/0	0/0	51/0	0/0	98/112
27	0/0	20/29	0/0	0/0	0/0	241/284

[a] Flocked process water was placed in two 55-gallon plastic-lined drums and sealed. DL-Lactic acid (0.2%) was added to each drum through a 2-in. screw cap. A control drum was analyzed without further additions. The active drum was inoculated with 500 mL of *Actinomyces viscosus dechlorini* that was reductively dechlorinating PCP. Samples were withdrawn at specified intervals with a sterile plastic pipe (1 in. × 4 ft) for PCP and intermediate analysis. Identification and quantification of PCP and intermediates were performed by Savannah Laboratories and Environmental Services, Inc. (Savannah, GA) for this table.

On-Site Biotransformation of PCP-Containing Flocked Process Water

Table 12.4 shows the results of an on-site treatment of flocked process water with *A. viscosus* sp. strain *dechlorini*. Two 55-gallon, plastic-lined drums were filled with flocked process water and sealed. A 2-in. screw cap port in the drum top was used to withdraw samples for PCP and intermediate product analysis. DL-Lactic acid (0.2%) was added through the port in each drum. The control drum was analyzed without further additions. To the active drum was added 500 mL of the bacterium that was actively dechlorinating PCP. Both drums transformed PCP by day 27. The active drum appeared to remove PCP at a faster rate through day 9. However, on day 14, the control drum had transformed more PCP than the active, 262 vs. 240 µg/mL, respectively. Intermediate product formation indicated the presence of 2,3,4,6-TeCP, 2,4,6-TCP, 2,4-DCP, 2-CP, and phenol, which was not consistent to that observed for *A. viscosus* sp. strain *dechlorini* in pure culture.

CONCLUSIONS

The biotransformation of PCP by *A. viscosus* strain *dechlorini* has an obligate requirement for yeast extract. This requirement may be the result of a requirement for an organic source of nitrogen, which is common in *Actinomyces* species.[43] Yeast extract may also provide cofactors or other components that facilitate reductive dechlorination and/or anaerobic respiration. The bacterium also demonstrates an improved capacity to biotransform greater concentrations of PCP when DL-lactic acid is added to the medium. This finding is consistent with the results of other workers that have shown that lactate and other short-chain organic acids often stimulate reductive dechlorination.[5,35,48] The organic acids apparently serve as electron donors to reduce chlorinated compounds and as a carbon and energy source for the bacterium. The results shown in Figures 12.3 and 12.4a and b suggest that DL-lactic acid serves a similar purpose for *A.*

viscosus strain *dechlorini*. By contrast, biotransformation of PCP by *A. viscosus* strain *dechlorini* was not enhanced by additions of glucose, ethanol, heat treatment, iron, succinate, or acetate as observed by other workers.[5,35,48–52] PCP serves as an electron acceptor, apparently increasing available cellular energy for bacterial growth/replication as indicated in Figure 12.4a. These results suggest a cocatabolic relationship between DL-lactic acid and PCP dechlorination indicative of a form of anaerobic respiration existing in *A. viscosus* strain *dechlorini*.

A. *viscosus* strain *dechlorini* biotransforms PCP under anaerobic conditions, but tetrachlorophenols do not accumulate in sufficient amount for identification. Instead, the first observed intermediate produced by the bacterium when grown on yeast extract and DL-lactic acid is 2,3,5-TCP. Without lactic acid, 2,4,5-TCP was the first observed intermediate. Variation in product formation due to medium composition has been observed by other workers.[35] A. *viscosus* strain *dechlorini* when grown on lactic acid and yeast extract under anaerobic conditions apparently follows the pathway of reductive dechlorination of PCP shown in Figure 12.8 as determined from Figures 12.3 and 12.4b. Recently, another group has observed that PCP is reduced to 2,3,5,6-TeCP and then to 2,3,5-TCP by a methanogenic consortium adapted to PCP.[36] However, 2,3,5-TCP was dechlorinated to 3,5-DCP while *A. viscosus* strain *dechlorini* produces 2,5-DCP. Significantly, *A. viscosus* strain *dechlorini* does not generate 2,3,4,5-TeCP, 2,3,4-, 3,4,5-, 2,4,5-, 2,4,6-, or 2,3,6-TCP, which have been shown to be toxic[51] and/or to induce DNA damage.[53] That is, of the six possible trichlorophenols that could be generated from PCP, the only one that has not been associated with being clastogenic, 2,3,5-TCP, is produced by the bacterium. Consequently, biotransformation of PCP by this bacterium apparently avoids the production of toxic or mutagenic intermediates entirely.

Attempted bioremediation of aqueous samples from Meredith Wood treatment facility using *A. viscosus* strain *dechlorini* resulted in an increase in the rate of PCP removal in the laboratory. However, the samples from the wood treatment facility were enriched with indigenous microorganisms capable of biotransformation of PCP since control samples removed PCP and other contaminants readily. Imposition of anaerobic conditions and supplementation with lactic acid apparently induced reductive dechlorination in the indigenous consortia. Figure 12.8 also shows the intermediate products of reductive dechlorination by indigenous microorganisms of the site samples. Each of the three samples followed the same dechlorination pattern but with minor exceptions. Unflocked process water did not accumulate 2,3,4,6-TeCP. Flocked process water and groundwater accumulated 2,4,5-TCP as well as 2,4,6-TCP. Groundwater accumulated 2,6-DCP as well as 2,4-DCP. Since many of these intermediates are toxic or mutagenic as mentioned above, the potential benefit of effecting biotransformation of PCP using *A. viscosus* strain *dechlorini* is apparent.

Actinomyces strains are capable of biodegrading cellulose, hemicellulose, lignocellulose, lignin, and starch components of agricultural and urban waste.[54] Based on the results of this study, this capability now extends to the anaerobic

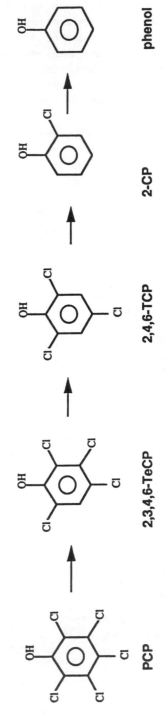

Figure 12.8   Apparent pathway for PCP biotransformation by *A. viscosus* sp. strain *dechlorini* and indigenous microorganism in unflocked, flocked, and groundwater from Meredith wood treatment facility.

biotransformation of chlorinated aromatic compounds. Continued search and development of *Actinomyces* strains having such capacities should prove of benefit to agribusiness and other industries given the continued use of such compounds as PCP, 2,4-D, 2,4,5-T, chlorinated anilines, and PCBs.

REFERENCES

1. Lee, M. D., Thomas, J. M., Borden, R. C., Bedient, P. B., and Ward, C. H. *Crit. Rev. Environ. Control.*, 18, 29, 1988.
2. Mikesell, M. D. and Boyd, S. A. *Appl. Environ. Microbiol.* 52, 861, 1985.
3. Mikesell, M. D. and Boyd, S. A. *Appl. Environ. Microbiol.* 53, 261, 1986.
4. Leahy, J. G. and Colwell, R. R. *Microbiol. Rev.* 54, 305, 1990.
5. Bryant, F. O., Hale, D. D., and Rogers, J. E. *Appl. Environ. Microbiol.* 57, 2293, 1991.
6. Gibson, S. A. and Sewell, G. W. *Appl. Environ. Microbiol.* 58, 1392, 1992.
7. Genthner, B. R. S. and Price, II, W. A., and Pritchard, P. H. *Appl. Environ. Microbiol.* 55, 1472, 1989.
8. Hale, D. D., Rogers, J. E., and Wiegel, J. *Microbial Ecol.* 20, 185, 1990.
9. Mohn, W. W. and Kennedy, K. J. *Appl. Environ. Microbiol.* 58, 1367, 1992.
10. Mueller, J. G., Middaugh, D. P., Lantz, S. E., and Chapman, P. J. *Appl. Environ. Microbiol.* 57, 1277, 1991.
11. Nishino, S. F., Spain, J. C., Belcher, L. A., and Litchfield, C. D. *Appl. Environ. Microbiol.* 58, 1719, 1992.
12. Struijs, J. and Rogers, J. E. *Appl. Environ. Microbiol.* 55, 2527, 1989.
13. Kuhn, E. P. and Suflita, J. M. *Reactions and Movements of Organic Chemicals in Soil.* Soil Science Society of America, 1989, Vol. 22, p. 111–180.
14. Guthrie, M. A., Kersch, E. J., Wukasch, R. F., and Grady, C. P. L. Jr. *Water Res.* 18, 451, 1984.
14a. Bauchinger, M., Dresp, J., Schmid, E., and Hauf, R. *Mutat. Res.* 102, 83, 1982.
15. Yokoyama, M. T., Johnson, D. A., and Gierzak, J. *Appl. Environ. Microbiol.* 54, 2619, 1988.
16. Crosby, D. G. *Pure Appl. Chem.* 53, 1050, 1981.
17. Paasivirta, J., Heinola, K., Humppi, T., Karjalainen, A., Knuutinen, K., Mantykoski, K., Paukku, R., Piilola, T., Surma-Aho, K., Tarhanen, J., Welling, L., Vihonen, H., and Sarkka, J. *Chemosphere.* 14, 469, 1985.
18. Keith, L. H. and Telliard, W. A. *Environ. Sci. Technol.* 13, 416, 1979.
19. Adams, R. H., Huang, C.-M., Higson, F. K., Brenner, V., and Focht, D. D. *Appl. Environ. Microbiol.* 58, 647, 1992.
20. Ahmed, M. and Focht, D. D. *Can. J. Microbiol.* 19, 47, 1973.
21. Ahmed, D., Sylvestre, M., and Sondossi, M. *Appl. Environ. Microbiol.* 57, 2880, 1991.
22. Copley, S. D. and Crooks, G. P. *Appl. Environ. Microbiol.* 58, 1385, 1992.
23. Kiyohara, H., Hatta, T., Ogawa, Y., Kakuda, T., Yokoyama, H., and Takizawa, N. *Appl. Environ. Microbiol.* 58, 1276, 1992.
24. Li, D.-Y., Eberspacher, J., Wagner, B., Kuntzer, J., and Lingens, F. *Appl. Environ. Microbiol.* 57, 1920, 1991.
25. Nishino, S. F., Spain, J. C., Belcher, L. A., and Litchfield, C. D. *Appl. Environ. Microbiol.* 58, 1719, 1992.

26. Stromo, K. E. and Crawford, R. L. *Appl. Environ. Microbiol.* 58, 727, 1992.
27. Topp, E., Xun, L., and Orser, C. S. *Appl. Environ. Microbiol.* 58, 502, 1992.
28. Valli, K., Wariishi, H., and Gold, M. H. *J. Bacteriol.* 174, 2131, 1992.
29. Lamar, R. T. and Dietrich, D. M. *Appl. Environ. Microbiol.* 56, 3093, 1990.
30. Niles, L. and Vogel, T. M. *Appl. Environ. Microbiol.* 57, 2771, 1991.
31. Allard, A.-S., Hynning, P.-A., Lindgren, C., Remberger, M., and Neilson, A. *Appl. Environ. Microbiol.* 57, 77, 1991.
32. Allard, A.-S., Hynning, P.-A., Remberger, M., and Neilson, A. *Appl. Environ. Microbiol.* 58, 961, 1992.
32a. Edwards, E. A. and Grbic-Galic, D. *Appl. Environ. Microbiol.* 58, 2663, 1992.
33. Larsen, S., Hendriksen H. V., and Ahring, B. K. *Appl. Environ. Microbiol.* 57, 2085, 1991.
34. Madsen, T. and Aamand, J. *Appl. Environ. Microbiol.* 57, 2453, 1991.
35. Mohn, W. W. and Kennedy, K. J. *Appl. Environ. Microbiol.* 58, 2131, 1992.
36. Nicholson, D. K., Woods, S. L., Istok, J. D., and Peek, D. C. *Appl. Environ. Microbiol.* 58, 2280, 1992.
37. Van Dort, H. M. and Bedard, D. L. *Appl. Environ. Microbiol.* 57, 1576, 1991.
38. DeWeerd, K. A., Mandelco, L., Tanner, R. S., Woese, C. R., and Suflita, J. M. *Arch Microbiol.* 154, 23, 1990.
39. DeWeerd, K. A., Concannon, F., and Suflita, J. M. *Appl. Environ. Microbiol.* 57, 1929, 1991.
40. Dolfing, J. and Tiedje, J. M. *Appl. Environ. Microbiol.* 57, 820, 1991.
41. Griffith, G. D., Cole, J. R., Quensen III, J. F., and Tiekje, J. M. *Appl. Environ. Microbiol.* 58, 409, 1992.
42. Curtis, T. *Texas Monthly* 112, 1990.
43. Gottlieb, D. *Bergey's Manual of Determinative Bacteriology,* 8th ed. Williams & Wilkins, Baltimore, 1974, pp. 657–681.
44. Smibert, R. M. and Krieg, N. R. *Manual of Methods or General Bacteriology.* American Society for Microbiology, Washington, D.C., 1981, pp. 409–443.
45. Cross, T. and Goodfellow, M. *Actinomycetales: Characteristics and Practical Importance.* Academic Press, New York, 1973, pp. 11–21.
46. Georg, L. K. and Gerencser, M. A. *Int. J. Syst. Bacteriol.* 19, 291, 1969.
47. Horvat, R. J., Chapman, G. W., Robertson, J. A., Meredith, F. I., Scorza, R., Callahan, A. M., and Morgens, P. *J. Agri. Food Chem.* 38, 234, 1990.
48. Holliger, C., Schraa, G., Stams, A. J. M., and Zehnder, A. J. B. *Appl. Environ. Microbiol.* 58, 1636, 1992.
49. Beller, H. R., Grbic-Galic, D., and Reinhard, M. *Appl. Environ. Microbiol.* 58, 786, 1992.
50. Hendriksen, H. V., Larsen, S., and Ahring, B. K. *Appl. Environ. Microbiol.* 58, 365, 1992.
51. Madsen, T. and Aamand, J. *Appl. Environ. Microbiol.* 58, 557, 1992.
52. Ye, D., Quensen III, J. F., Tiekje, J. M., and Boyd, S. A. *Appl. Environ. Microbiol.* 58, 1110, 1992.
53. DeMarini, D. M., Brooks, H. G., and Parkes, D. G. Jr. *Environ. Molec. Mutagen.* 15, 1, 1990.
54. Crawford, D. L. *Actinomyces in Biotechnology.* Academic Press, San Diego, 1988, pp. 433–450.

CHAPTER 13

A Versatile Bioluminescent Reporter System for Organic Pollutant Bioavailability and Biodegradation

A. Heitzer, O. F. Webb, P. M. DiGrazia, and Gary S. Sayler

INTRODUCTION

The industrial development of our society and the concomitant population growth together with the resulting increased demand for natural resources have resulted in a significant local but also diffuse global pollution of our environment with a large variety of pollutants. During the last decade or two, governmental efforts were directed at controlling and preventing further pollution by implementing new legal guidelines for production processes as well as at the clean-up of existing pollution in various different environments involving soils, aquifers, sediments, and surface water. Programs to evaluate and promote new, innovative treatment technologies have emerged since and national priority sites for clean-up were defined in the superfund act. A number of possible treatment technologies exist that involve physical, chemical, and biological methods or combinations thereof; however, prior to any successful use of these methods, a site needs to be characterized and evaluated. The characterization of a contaminated site requires an integrated, interdisciplinary approach involving physicochemical, hydrogeological, and biological characterizations, which together with legal and economical consideration have to be used for an evaluation of appropriate treatment scenarios.

For the biological characterization of a site, a number of traditional but also novel techniques involving recent molecular biological methodologies have become available. The development of a number of bioanalytical methods include molecular diagnostic techniques that comprise application of gene probe techniques for analysis of specific genotypes and their activities,[1-4] PCR technology,[5] lipid-fatty acid analysis,[6] as well as the use of monoclonal antibodies for the specific detection of bacterial species in natural environments.[7] Most recently, genetically engineered bacteria carrying gene fusions between cata-

0-87371-951-4/95/$0.00+$.50

bolic genes and the genes encoding for bacterial bioluminescence have been developed and used as bioanalytical tools to assess pollutant bioavailability and biodegradation.[8–10] Other conventional methods involve the use of mineralization assays for biodegradability and bioavailability studies or material balance-based biodegradation studies.[11]

An important point to be taken into account in the development of any biological method to be used for environmental studies is the fact that real polluted sites are complex, dynamic, heterogeneous environments that involve microbial communities and mixtures of pollutants rather than monocultures, single substrates, and constant environmental conditions.[12] The pollutants present might be degradable or recalcitrant and might often exhibit toxic effects on indigenous or introduced bacteria that might significantly affect their activities and efficacies.

It is the goal of this contribution to discuss the principles of the catabolic bioluminescent reporter technology based on the example of a reporter bacterium for naphthalene and salicylate metabolism. Similar systems have been developed for the detection of heavy metals.[13] The versatility of this approach is highlighted with several examples. Some potential applications or areas for applications in environmental biotechnology are indicated.

BIOLUMINESCENT CATABOLIC REPORTER BACTERIA: RATIONALE AND STRAIN CONSTRUCTION

An important issue for the successful prediction of microbial degradation of nonpolar organic pollutants such as polyaromatic hydrocarbons (PAH) in contaminated environments concerns their bioavailability and relative biodegradability. In complex, heterogeneous environments such as soils and sediments, the relative distribution of these compounds among the solid, aqueous, and gaseous phases determines whether the pollutant is accessible to the bacteria. This behavior is largely due to low aqueous solubilities and to the high tendency of these compounds to adsorb to soil or sediment particles.

The bacterial degradation of a number of pollutants is mediated by inducible catabolic operons that are often induced by the pollutant itself or by one of its degradation intermediates. In many cases these operons are plasmid encoded.[14] The degradation of the priority pollutant naphthalene, as encoded on the NAH7 plasmid, is mediated by two catabolic operons (Fig. 13.1). The enzymes encoded by the naphthalene (upper) operon catalyze the degradation of naphthalene to salicylate, while the enzymes encoded by the salicylate (lower) operon are responsible for the conversion of salicylate to 2-oxo-4-hydroxypentanoate. The latter compound is then further metabolized to pyruvate and acetaldehyde, which both enter the basic metabolic pathways of the bacterium.[15]

A regulatory gene, nahR, is located between the two operons. The nahR gene product binds to the DNA in the promoter regions of the two operons and

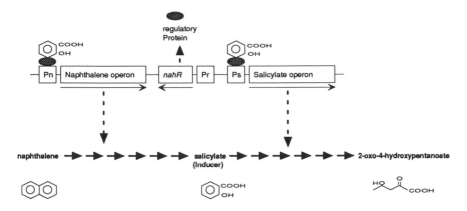

Figure 13.1 Organization and regulation of the naphthalene operon on the NAH7 plasmid. Thin arrows indicate direction of transcription, dashed arrows symbolize transcription and translation processes, and thick arrows represent enzyme reactions. Pn, Ps, and Pr represent promoters.

activates transcription in interaction with the naphthalene degradation interme-diate, salicylate, which functions as the inducer of the naphthalene operon.[16] Therefore, exposure of the bacterium to either naphthalene or salicylate results in simultaneously increased expression of the two operons. The expression of such catabolic operons provides the basis to assess the bioavailability of a certain pollutant or a group of pollutants since an increased expression of a catabolic operon indicates that a compound could enter the bacterial cell.

A number of methods exist for the measurement of gene expression.[1] Many of these techniques require the disintegration of the bacterial cell and involve a significant amount of work and time until results are obtained. It was the genetic characterization of the *lux* system in the bioluminescent bacterium *Vibrio fischeri*[17] and the subsequent demonstration that these genes can be used as reporters in gene fusions to monitor gene expression simply by measuring emitted light[18] that has opened a new field for applications of this system in basic and applied molecular biology. The novel feature of this approach to measure gene expression was that it allows *in situ*, real-time analysis of gene expression without the need for cell disruption.

During the last two decades the ecology,[19] biochemistry, and physiology,[20,21] as well as regulation and genetics[22,23] of bacterial bioluminescence, were inves-tigated in detail. The bioluminescence reaction is catalyzed by a heterodimeric luciferase, which is encoded by the *luxA* and *luxB* genes. The light reaction itself is dependent on a substrate fatty aldehyde, $FMNH_2$ and O_2, which are con-verted to a fatty acid, FMN and H_2O, and, as a by-product, light is emitted. The synthesis and recycling of the fatty aldehyde from fatty acids are catalyzed in an ATP- and NADPH-dependent manner by a multienzyme fatty acid reductase complex, comprising a reductase, a transferase, and a synthetase, which are encoded by the *luxC*, *luxD*, and *luxE* genes, respectively. For the use of the *lux* genes as reporters for gene expression, two strategies might be used: one

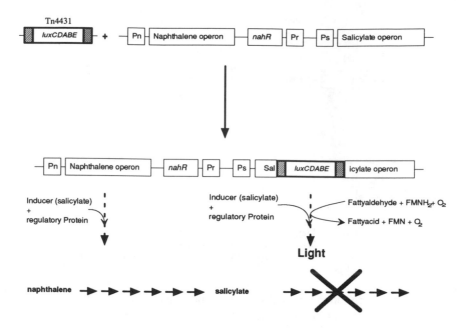

Figure 13.2   Construction of a catabolic naphthalene–*lux* reporter gene fusion using transposon mutagenesis.

involving the *luxAB* genes and another one involving the *luxCDABE* genes. While the former system makes an addition of the aldehyde substrate necessary, the latter would be independent of such an addition in a compatible host bacterium. The choice of the source bacterium of the *lux* genes is largely dependent on the planned application of the reporter bacterium. Advantages and disadvantages of different *lux* systems are discussed by Meighen.[23]

Recently, two different constructions of bioluminescent reporter bacteria for naphthalene and salicylate catabolism have been reported.[8,9] Figure 13.2 outlines a transposon mutagenesis strategy that might be used when no detailed information on restriction sites for the genes of interest is available. This approach was chosen by King et al.[9] because their bacterium of interest was a naphthalene degrading environmental isolate, designated as *P. fluorescens* 5R, from a contaminated manufactured gas plant soil. Such environmental isolates from real waste sites might be particularly robust and, therefore, well suited for environmental applications.

It was demonstrated, using DNA hybridization techniques, that the genes encoding naphthalene degradation were located on a plasmid, pKA1, that exhibited structural homology to the archetypal NAH7 plasmid. This isolate was then subjected to conjugation with *E. coli* HB101, which harbored the *luxCDABE* genes containing transposon Tn*4431* on a suicide plasmid pUCD623.[24] A transconjugant, *P. fluorescens* 5RL, was subsequently found that had the transposon inserted into the *nahG* gene of the salicylate operon. The plasmid containing the insertion was named pUTK21. This strain produces light upon exposure to either naphthalene or salicylate; however, salicylate could not be further degraded in this strain because of the insertion of the *luxCDABE* gene cassette in the salicylate operon. Therefore, plasmid pUTK21 was transferred into another environmental isolate from the same site, capable of degrading salicylate but not naphthalene. The resulting strain, *P. fluorescens* HK44, produces light after exposure to naphthalene and salicylate and is also able to degrade both compounds.

BIOLUMINESCENT MONITORING OF DYNAMIC PERTURBATIONS ON CATABOLIC ACTIVITIES

For the evaluation of the efficiency of a bioremediation process, comparative performance studies of the system under a wide range of realistic environmental conditions have to be conducted. Of particular interest are dynamic responses of a process culture to perturbation to obtain information on the robustness of a particular system and also to define its efficiency and performance ranges. Different scenarios can be envisioned for such experiments; depending on a process design, studies can be conducted in batch or continuous cultures. The reporter bacteria might be applied as pure cultures of one principal biodegradative bacterium or as addition to an already existing mixed culture.

While the former scenario can provide valuable information about the effects of environmental factors on overall process performance as well as on physiological aspects such as specific gene expression and catabolic activity of a particular bacterium, the latter would, in addition, provide information on the stability and persistence of an introduced bacterium in a mixed culture.

DiGrazia[25] has investigated the effects of naphthalene feed perturbations on catabolic gene expression and naphthalene degradation in a continuous culture of the bioluminescent reporter bacterium *P. fluorescens* HK44. For this study, a frequency–response analysis was adopted that involves the variation of the input variable, naphthalene concentration, in a sinusoidal manner with specific amplitude and cycle frequency. In the perturbed system, the output signal, bioluminescence, was monitored and an analysis was conducted using Fourier transformations to separate the sinusoidal output signal from the noise. Experiments were performed at various different cycle frequencies to obtain information on the systems response to perturbations over a range of time scales. In general, this method is applicable for system response analysis to perturbations in any important variable.

The culture was grown in a mineral salts medium containing as carbon substrates constant amounts of 100 mg/L sodium succinate and 100 mg/L yeast extract and a varying concentration of naphthalene from 0 to 30 mg/L. The bioreactor was operated in a continuous mode at pH 7 and 25°C with a stirring rate of 500 rpm and an air flow rate of 6 L/hr. The reactor volume and feed flow rate were 0.75 L and 0.3 L/hr, respectively. Bioluminescence was measured

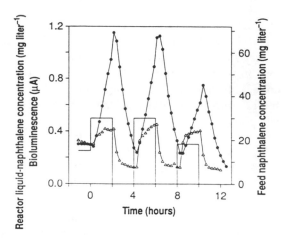

Figure 13.3   Dynamic bioluminescence response of *Pseudomonas fluorescens* HK44 in continuous culture to 4-hr square-wave perturbations in the naphthalene feed concentration: (●) culture bioluminescence, (—) naphthalene feed concentration, (Δ) reactor liquid naphthalene concentration. From King et al.[9]

with a liquid light tube in conjunction with a photomultiplier/recorder unit and data were stored on-line with a computer. The reactor liquid naphthalene concentration was determined by gas chromatography based on the Henry's law coefficient using head space gas-phase measurements as described by DiGrazia et al.[26]

In Figure 13.3 the response of the bioluminescent reporter bacterium *P. fluorescens* HK44 to square wave perturbations in the naphthalene feed is shown. The reactor system used allowed for variations in the feed naphthalene concentrations under a constant feed flow rate.[26] The experiments involved a 4-hr cycle. Increase in naphthalene feed concentration resulted in both increased residual naphthalene concentration and increased bioluminescence, which continued at a constant rate throughout the 2-hr period where the reactor feed contained naphthalene. Absence of naphthalene in the reactor feed resulted in a decrease in the residual naphthalene concentration and simultaneous decrease in the bioluminescence response.

This experiment demonstrates the positive correlation between reactor liquid naphthalene concentration and bioluminescence. Subsequently, the same system was subjected to sinusoidal perturbations in the naphthalene feed ranging from 1- to 12-hr cycle periods. Using a frequency–response analysis, it was possible to determine critical frequency ranges as shown in the Bode diagrams in Figure 13.4.

In Figure 13.4A it is demonstrated that a critical frequency range exists between 0.25 and 0.5 cycles per hour, above which the amplitude ratio of the bioluminescence response to the reactor naphthalene concentration decreased. Figure 13.4B shows that the response is forced out of phase between 0.125 and 0.25 cycles per hour. Similarly, DiGrazia[25] had demonstrated that the amplitude

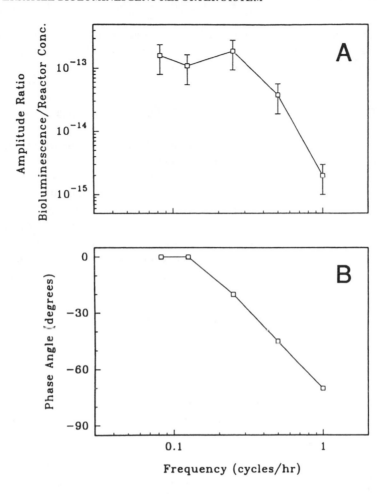

Figure 13.4   Bode diagram relating bioluminescence to reactor liquid naphthalene concentration
in continuous cultures of *P. fluorescens* HK44. (A) Amplitude ratio. (B) Phase angle.
From DiGrazia.[25]

ratio between reactor naphthalene concentration and naphthalene degradation
rate also had its critical range between 0.25 and 0.5 cycles per hour. However,
no phase shift was observed. These findings indicate that the bioluminescence
response is related to the biodegradation kinetics.

   This example illustrates the utility of bioluminescent reporter bacteria for
pollutant catabolism monitoring in dynamic systems under defined conditions
and also determines critical regimes for the use of the bioluminescence signal
as indicator for the reactor naphthalene concentration. The identification of
such critical regimes is particularly important for two reasons: first, the robustness of a bioprocess to environmental perturbations can be determined, and
application ranges for bioluminescent reporter bacteria may be defined. Sec-

ond, basic information can be obtained on the response of bacteria and their catabolic systems to environmental perturbations, which contributes to a fundamental understanding of bacterial ecophysiology.

QUANTIFICATION OF POLLUTANT BIODEGRADATION

From an applied point of view, the use of bioluminescent reporter bacteria for the estimation of pollutant biodegradation would be very useful to achieve rapid information on the status of process. Such information, in combination with chemical analytical data could provide useful and rapid insight into different treatment procedures and their relative efficacies on a comparative basis. Since it was demonstrated that bioluminescence and naphthalene degradation were coinciding processes under a wide range of cultivation conditions for the reporter bacteria,[8,9,25] it was of interest to determine whether the bioluminescence response could be used to estimate the amount of pollutant that has been degraded.

A batch culture was grown in 300-mL Erlenmeyer flasks at 26°C and pH 7, using a mineral salts medium buffered with 0.05 M phosphate. The medium contained 1 g/L glucose and 5.83 mg/L salicylate as carbon substrates. Biomass and salicylate were both measured spectrophotometrically at 546 and 296 nm, respectively. Figure 13.5A and B shows the time course of the specific culture bioluminescence (bioluminescence per biomass, measured as optical density at 546 nm) and the medium salicylate concentration in the culture filtrate, respectively. The bioluminescence signal exhibited a maximum at 1.25 hr. This experiment reflects an initial increase in gene expression, followed by a subsequent reduction in response to the continuously decreasing salicylate liquid concentration. In Figure 13.5C the total bioluminescence area for each time point taken was determined by numerical integration and plotted against the amount of salicylate that was degraded at the corresponding time points. A positive correlation was found that exhibited good linearity at this salicylate concentration.

Similar results were obtained under identical conditions for salicylate concentrations ranging from 3 to 12 mg/L. These results demonstrate that under defined, controlled conditions the specific bioluminescence response of the reporter culture provides a potentially valuable method for on-line estimation of pollutant degradation and process control; however, further experiments are necessary to investigate these relationships over a larger target compound concentration range and in complex environmental matrices.

DEVELOPMENT OF A BIOASSAY FOR POLLUTANT PRESENCE AND BIOAVAILABILITY MONITORING

Based on these previous findings, a bioassay was developed for the assessment of naphthalene and salicylate distribution and bioavailability in environ-

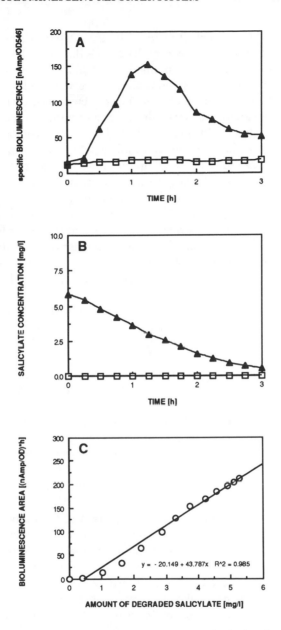

Figure 13.5   Relationship between specific bioluminescence and salicylate degradation in batch cultures of *P. fluorescens* HK44. (A) Time course of the specific bioluminescence response. (□) Control without salicylate; (▲) 5.83 mg/L salicylate. (B) Time course of the salicylate concentration. (□) Control without salicylate; (▲) 5.83 mg/L salicylate. (C) Integrated bioluminescence for each time point plotted against the amount of degraded salicylate at corresponding time points.

mental systems.[10] An important factor to be considered in any environmental application of bioluminescent reporter bacteria is that "real" contaminated sites are complex environments that commonly contain heterogeneous mixtures of pollutants, of which some compounds might be readily biodegradable, recalcitrant, or exhibit toxic effects on the reporter bacteria. All these factors could affect the performance of the test culture upon exposure to such matrices and exert a significant stress to the bacteria. In addition, the physiological state of the culture itself, when added to the test sample, might be important to be considered. Possible consequences of a variety of environmental stresses on the bacterial physiology have been discussed by Hamer and Heitzer,[12] and the concept of a generalized stress cycle has been introduced.

The assay was conducted in 25-mL mineralization flasks to keep the volume small, which would allow for easy handling, be potentially useful for field on-site applications, and make the simultaneous processing of a number of samples possible. The setup consisted of a light-tight measurement chamber with an inserted liquid light tube that is connected with a photomultiplier–recorder unit.

Initially, the possible impact of the physiological state of the reporter culture on the response to readily utilizable, but noninducing, carbon substrates was investigated. The responses of resting, carbon-starved cultures and exponentially growing cultures, when exposed to complex nutrients such as diluted LB medium or yeast extract peptone medium, were compared. Exponentially growing cultures were prepared in a phosphate-buffered (0.05 M, pH 7.0) yeast-extract–peptone–glucose medium (YEPG) at 27°C. At a predefined optical density of 0.35 at 546 nm, 2-mL aliquots of this culture suspension were added directly to 2 mL of a test solution. Resting cultures were prepared from an identically grown exponential culture, centrifuged, resuspended in a buffered mineral salts medium without carbon substrate, and incubated on a shaker at 27°C for 3 hr. Two-milliliter aliquots of this culture were added to 2-mL test sample.

In both cases, the bioluminescence response was then monitored in intervals of 15 min and compared to a control without any inducer and to one containing 3.45 mg/L salicylate. The specific bioluminescence responses (bioluminescence per biomass) after 1 hr incubation time are presented in Table 13.1. It is obvious that exposure of resting cultures to noninducing, readily metabolizable carbon substrates results in a significant, nonspecific bioluminescence response as compared to an aqueous control and also to the sample containing 3.45 mg/L salicylate.

A plausible explanation for this behavior found with resting cultures is that exposure to carbon substrates results in growth, and that growing bacteria exhibit a low constitutive expression of the naphthalene operon.[27] Such constitutive expression is reflected in the low but significant bioluminescence response of the exponentially growing control culture. In addition, the dependence of the bioluminescence reaction on a number of cellular compounds such as reduction equivalents and fatty aldehyde, which might not be abundant in

Table 13.1 Effect of the Physiological State of the *P. fluorescens* HK44 Reporter Culture on Specific Bioluminescence after Exposure to Inducing and Noninducing Carbon Substrates

	Specific bioluminescence after 1 hr (nA/OD 546)	
	Carbon starved, resting culture	Exponentially growing culture
Control[a]	0.09 ± 0.04	3.64 ± 0.05
YEP 1:4[b]	1.50 ± 0.10	3.36 ± 0.11
LB 1:8[c]	2.12 ± 0.19	3.32 ± 0.07
Salicylate[d]	11.36 ± 1.85	30.26 ± 0.95

[a] No carbon substrate.

[b] 50 mg/L yeast extract and 500 mg/L polypetone.

[c] 1.25 g/L tryptone, 0.625 g/L yeast extract and 1.25 g/L NaCl.

[d] 3.45 mg/L.

Adapted from Heitzer et al.[10]

resting, carbon-starved cultures, could affect the bioluminescence reaction directly. These results clearly demonstrate how important it is to take the physiological state of the test culture into account in any development of a whole cell-based bioassay for environmental applications. It was, therefore, decided to conduct further experiments with exponentially growing cultures.

Another important issue with respect to practical applications are effects of environmentally relevant potentially toxic pollutants. The effect of exposure to 1 mM cadmium chloride ($CdCl_2$), 1 mM mercury chloride ($HgCl_2$), and 1.32 mM potassium cyanide (KCN) solutions was investigated. It was found that all three compounds at the concentrations used caused an immediate reduction in bioluminescence compared to a control in water, which exhibits a low, basal bioluminescence. These results indicate that toxic compounds could interfere with this assay. However, the comparison to an aqueous control sample allows identification of significant inhibitory effects that result in a decrease in biolu-minescence below the noninduced basal level, analogous to existing biolumines-cence-based toxicity assays.[28]

The correlation between bioluminescence and initial naphthalene concen-tration is shown in Figure 13.6. Aliquots of an exponentially growing reporter culture were added, as described above, to test samples of different naphthalene concentrations and were incubated for 1 hr before the bioluminescence was determined. Bioluminescence was plotted against analytically determined ini-tial naphthalene concentration. A good linearity was found over a naphthalene concentration range of about 2 orders of magnitude. A significant biolumines-cence response was detectable at naphthalene concentrations as low as 45 ppb. Similar results were also obtained for salicylate.[10]

Subsequently, the complexity of the test system was increased to obtain information on the utility of this method to estimate concentrations of bioavailable pollutants in extracts and slurries of experimentally contaminated soils. A com-parative study using salicylate and naphthalene contaminated soils was con-ducted.[10] The soils had received either 6.9 mg/kg salicylate or 8.3 mg/kg naph-

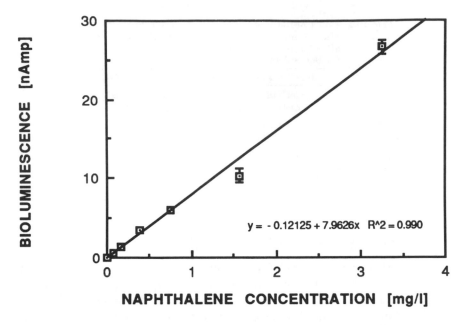

Figure 13.6   Relationship between initial naphthalene concentration and bioluminescence in exponentially growing *Pseudomonas fluorescens* HK44, after 1 hr incubation. Bioluminescence values are corrected for the response observed in the control without naphthalene. From Heitzer et al.[10]

thalene. Slurries were prepared using a one-to-one ratio of soil to water (w/v). The supernatants of these slurries, after centrifugation in Teflon-sealed corex glass tubes, provided the extracts. Aqueous solutions containing either 6.9 mg/L salicylate or 8.3 mg/L naphthalene were prepared together with the soils and used as standards for an estimation of pollutant concentration in the soil extract and slurry samples.

For all the experiments, uncontaminated, identically treated samples were included. The incubation period for all the samples was 1 hr. The results of this experiment are summarized in Table 13.2. In soil extracts, estimated pollutant concentrations were reasonably close to the analytically determined concentrations; however, in soil slurries no such estimations could be conducted because of significant light quenching effects by soil particles. The relative time course of the bioluminescence under these conditions was similar to those observed in soil extracts.[10] For both compounds, salicylate and naphthalene, the quenching effect was about 1 order of magnitude. If such a quenching factor could be accurately determined, quantitative estimates of pollutants in soil slurries might become possible.

The experiment was conducted so that the dissolved pollutant concentrations in all the salicylate and all the naphthalene samples could be theoretically the same provided no adsorption to the soil matrix takes place. In the case of salicylate, the bioluminescence response and the analytically determined con-

Table 13.2 Estimation of Bioavailable Naphthalene and Salicylate in Soil Extracts and Soil Slurries Using *P. fluorescens* HK44

	Salicylate			Naphthalene		
	Bioluminescence after 1 hr[a] (nA)	Estimated initial concentration (mg/L)	Analytically determined initial concentration (mg/L)	Bioluminescence after 1 hr[a] (nA)	Estimated initial concentration (mg/L)	Analytically determined initial concentration (mg/L)
Standard	11.05 ± 0.4	—	3.45 ± 0.07	55.71 ± 3.4	—	4.15 ± 0.36
Soil extract	12.28 ± 1.0	3.83 ± 0.32	2.90 ± 0.12	2.41 ± 0.1	0.18 ± 0.02	0.16 ± 0.05
Soil slurry	0.82 ± 0.4	[b]	n.d.[c]	0.27 ± 0.02	[b]	n.d.

[a] The bioluminescence values are corrected for the response observed in the uncontaminated controls.

[b] No estimation possible due to light quenching effects by soil particles.

[c] n.d., not determined.

Adapted from Heitzer et al.[10]

centrations corresponded to the aqueous test sample, indicating that no signifi-cant adsorption to the soil matrix took place. In contrast, a significant discrep-ancy was found between test sample and soil extract with naphthalene, which indicates that a significant portion of naphthalene had adsorbed to the soil matrix.

These experiments have demonstrated the feasibility of using biolumines-cent reporter bacteria in complex environmental matrices to assess specifically and quantitatively pollutant bioavailability. Possible limitations have been indi-cated, such as the light-quenching effects in soil slurries as well as the possible interference of toxic compounds. It was further pointed out that the physiologi-cal state of the reporter culture is an important consideration.

For environmental applications, such an assay could prove useful to rapidly locate pollution plumes by assaying a number of samples from different depths and different locations at a contaminated site. Further, the relative efficiency of different procedures to desorb pollutants from contaminated soils could be rapidly assessed based on the bioluminescence response in comparative treat-ment studies.

BIOSENSORS

A novel application of bioluminescent reporter bacteria is their use for the development of a whole cell biosensor. A number of whole cell biosensors based on respirometric or calorimetric measurements have been reported for the monitoring of chemical compounds such as dichloromethane[29] and various aromatic compounds,[30,31] as well as for the determination of bulk parameters such as BOD.[32] The use of bioluminescent reporter bacteria provides a basis for the development of a new type of whole cell biosensor based on the light emission of the bacteria upon exposure to a pollutant.

Previous experiments have shown that alginate provides a clear, translucent immobilizing matrix for bioluminescent reporter bacteria, which allows light to pass through. Using bacteria entrapped in alginate beads, we have found a substrate concentration-dependent bioluminescence response to salicylate and naphthalene. Based on these findings, a biosensor was designed around com-mercially available light measurement equipment. In Figure 13.7A, a scheme of the biosensor setup is depicted. The biosensor includes the reporter bacteria, immobilized at the tip of a liquid light guide, which was connected to a photo-multiplier/recorder unit. Data were stored on-line using a personal computer and a software program (Bunnware, University of Tennessee). Raw data can be easily transferred to a spreadsheet for further processing. The probe tip was inserted into a thermostated vial on a stirrer. A scheme of the biosensor probe tip is shown in Figure 13.7B. A stainless-steel ferrule with a spherical stainless-steel mesh opening (109 μm) on one side was mounted on the tip of the liquid light guide. The immobilized bioluminescent reporter bacterium *P. fluorescens* HK44 was contained between the surface of the liquid light probe and the stainless-steel mesh.

A

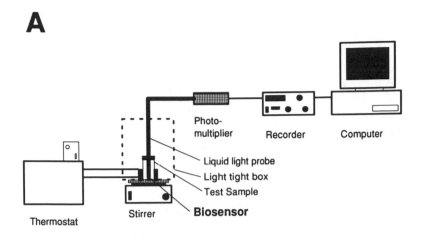

B

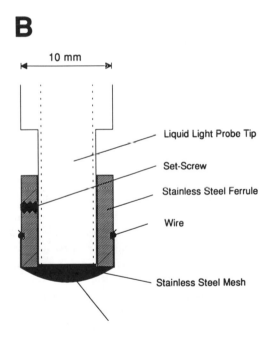

Figure 13.7   (A) Biosensor setup. (B) Biosensor probe tip.

In Figure 13.8, the response of this biosensor to an aqueous solution of 2.5 mg/L salicylate is shown and compared to a control of water. The experiment was conducted at neutral pH. Similar results were also obtained in slurries of soils contaminated with naphthalene. Although these results are preliminary and further experimentation is needed, such a system could find a wide range of applications in the laboratory analysis of environmental samples and potentially also in the field for *in situ* applications. In a continuous application, such a setup might also prove useful for the on-line monitoring of waste streams.

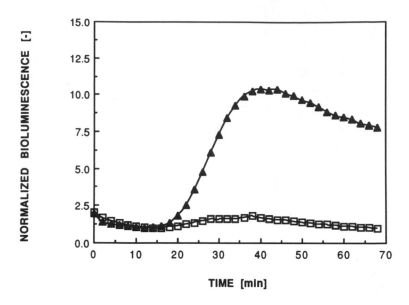

Figure 13.8   Response of the biosensor containing *P. fluorescens* HK44 to 2.5 mg/L salicylate (▲) and to water (□).

SUMMARY AND CONCLUSIONS

Bioluminescent reporter bacteria are suitable tools for the investigation of dynamic environmental changes on catabolic gene expression and bacterial physiology under *in situ* and real-time conditions.

This technology provides a valuable method for the rapid, specific, and quantitative assessment of pollutant bioavailability and biodegradation in simple assays that have a high potential for applications in "real" environmental systems; however, for the use of bioluminescent reporter bacteria in environmental applications, it is important to take the complexity of such "real systems" and the potentially resulting physiological effects on the performance of the reporter bacteria into account.

Experiments have demonstrated that it is feasible to use bioluminescent reporter bacteria for the development of whole cell biosensors. In principle, the bioluminescent reporter technology is of general applicability for any inducible genetic system. In combination with other bioanalytical and chemical analytical methods, it provides a valuable tool for the biological characterization of contaminated sites. It will facilitate and help in optimizing environmental applications of bacteria and their implementation in remediation processes.

ACKNOWLEDGMENTS

This work was supported by the U.S. Air Force Office of Scientific Research, Contract F49620-89-C-0023, and by the U.S. Department of Energy, Office of

Energy Research Subsurface Science Program Grant DE FG05-91ER61193. The technical assistance of Janeen Thonnard and Taffany Jones was greatly appreciated.

REFERENCES

1. Jain, R. K., Burlage, R. S., and Sayler, G. S. Methods for detecting recombinant DNA in the environment. *Crit. Rev. Biotechnol.* 8, 33, 1988.
2. Sayler, G. S. and Layton, A. C. Environmental applications of nucleic acid hybridization. *Annu. Rev. Microbiol.* 44, 625, 1990.
3. Sayler, G. S., Nikbakht, K., Fleming, J. T., and Packard, J. Applications of molecular techniques to soil biochemistry. In *Soil Biochemistry*, Vol. 7. Stotzky, G. and Bollag, J. M., Eds. Marcel Dekker, New York, 1992, p. 131.
4. Stahl, D. A. and Amann, R. Development and applications of nucleic acids probes. *Nucleic Acid Techniques in Bacterial Systematics*. Stackebrandt, E. and Goodfellow, M., Eds. John Wiley, New York, 1991, P. 205.
5. Atlas, R. M., Sayler, G. S., Burlage, R. S., and Bej, A. K. Molecular approaches for environmental monitoring of microorganisms. *BioTechniques* 12, 706, 1992.
6. Geesey, G. G. and White, D. C. Determination of bacterial growth and activity at solid–liquid interfaces. *Annu. Rev. Microbiol.* 44, 579, 1990.
7. Bohlool, B. B. and Schmidt, E. L. The immunofluorescence approach in microbial ecology. *Adv. Microb. Ecol.* 4, 203, 1980.
8. Burlage, R. S., Sayler, G. S., and Larimer, F. Monitoring of naphthalene catabolism by bioluminescence with *nah-lux* transcriptional fusions. *J. Bacteriol.* 172, 4749, 1990.
9. King, J. M. H., DiGrazia, P. M., Applegate, B., Burlage, R., Sanseverino, J., Dunbar, P., Larimer, F., and Sayler, G. S. Rapid, sensitive bioluminescent reporter technology for naphthalene exposure and biodegradation. *Science* 249, 778, 1990.
10. Heitzer, A., Webb, O. F., Thonnard, J. E., and Sayler, G. S. Specific and quantitative assessment of naphthalene and salicylate bioavailability using a bioluminescent catabolic reporter bacterium. *Appl. Environ. Microbiol.* 58, 1839, 1992.
11. Rochkind, M. L., Blackburn, J. W., and Sayler, G. S. Microbial decomposition of chlorinated aromatic compounds. *United States Environmental Protection Agency Report,* EPA/600/2-86/090, 269, 1986.
12. Hamer, G. and Heitzer, A. Polluted heterogeneous environments: Macro-scale fluxes, micro-scale mechanisms, and molecular scale control. *Environmental Biotechnology for Waste Treatment.* Sayler, G. S. et al., Eds. Plenum Press, New York, 1991, p. 233.
13. Taylor, J., Frackman, S., Langlay, K. M., and Rosson, R. A. Luminescent biosensors for detection of inorganic and organic mercury. ASM abstract No. Q323, American Society for Microbiology, 92nd Annual Meeting, New Orleans, LA, 1992.
14. Sayler, G. S., Hooper, S. W., Layton, A. C., and King, J. M. H. Catabolic plasmids of environmental and ecological significance. *Microb. Ecol.* 19, 1, 1990.
15. Yen, K. M. and Serdar, C. M. Genetics of naphthalene catabolism in pseudomonads. *Crit. Rev. Microbiol.* 15, 247, 1988.

16. Schell, M. A. Regulation of the naphthalene degradation genes of plasmid NAH7: Example of a generalized positive control system in *Pseudomonas* and related bacteria. *Pseudomonas: Biotransformations, Pathogenesis and Evolving Biotechnology.* Silver, S., Chakrabarty, A. M., Iglewski, B., and Kaplan, S., Eds. American Society for Microbiology, Washington, D.C., 1990, p. 165.

17. Engebrecht, J., Nealson, K. H., and Silverman, M. Bacterial bioluminescence: Isolation and genetic analysis of functions from *Vibrio fischeri*. *Cell* 32, 773, 1983.

18. Engebrecht, J., Simon, M., and Silverman, M. Measuring gene expression with light. *Science* 227, 1345, 1985.

19. Nealson, K. H. and Hastings, J. W. Bacterial bioluminescence: Its control and ecological significance. *Microbiol. Rev.* 43, 496, 1979.

20. Hastings, J. W. and Nealson, K. H. Bacterial bioluminescence. *Annu. Rev. Microbiol.* 31, 549, 1977.

21. Hastings, J. W., Potrikus, C. J., Gupta, S. C., Kurfürst, M., and Makemson, J. C. Biochemistry and physiology of bioluminescent bacteria. *Adv. Microb. Physiol.* 26, 235, 1985.

22. Meighen, E. A. Enzymes and genes form the *lux* operons of bioluminescent bacteria. *Annu. Rev. Microbiol.* 42, 151, 1988.

23. Meighen, E. A. Molecular biology of bacterial bioluminescence. *Microbiol. Rev.* 55, 123, 1991.

24. Shaw, J. J., Settles, L. G., and Kado, C.I. Transposon Tn*4431* mutagenesis of *Xanthomonas campestris* pv. *campestris:* Characterization of a nonpathogenic mutant and cloning of a locus for pathogenicity. *Mol. Plant-Microbe Interact.* 1, 39, 1988.

25. DiGrazia, P. M. Microbial systems analysis of naphthalene degradation in a continuous flow soil slurry reactor. Ph.D. thesis, University of Tennessee, Knoxville, 1991.

26. DiGrazia, P. M., King, J. M. H., Blackburn, J. W., Applegate, B. A., Bienkowski, P. R., Hilton, B. L., and Sayler, G. S. Dynamic response of naphthalene biodegradation in a continuous flow soil slurry reactor. *Biodegradation* 2, 81, 1991.

27. Schell, M. A. Transcriptional control of the *nah* and *sal* hydrocarbon-degrading operons by the *nahR* gene product. *Gene* 36, 301, 1985.

28. Bulich, A. A. Bioluminescence assays. *Toxicity Testing Using Microorganisms,* Vol. 1. Bitton, G. and Dutka, B. J., Eds. CRC Press, Boca Raton, FL, 1986, p. 57.

29. Henrysson, T. and Mattiasson, B. A dichloromethane sensitive biosensor based on immobilized *Hyphomicrobium* DM2 cells. *International Symposium on Environmental Biotechnology.* Verachtert, H. and Verstraete, W., Eds. Royal Flemish Society of Engineers, Ostend, Belgium, 1991, p. 73.

30. Riedel, K. A., Naumov, A. V., Boronin, A. M., Golovleva, L. A., Stein, H. J., and Scheller, F. Microbial sensors for determination of aromatics and their chloroderivatives. I. Determination of 3-chlorobenzoate using a *Pseudomonas* containing biosensor. *Appl. Microbiol. Biotechnol.* 35, 559, 1991.

31. Thavarungkul, P., Hakanson, H., and Mattiasson, B. Comparative study of cell-based biosensors using *Pseudomonas cepacia* for monitoring aromatic compounds. *Anal. Chim. Acta* 249, 17, 1991.

32. Karube, I., Matsunaga, T., Mitsuda, S., and Suzuki, S. Microbial electrode BOD sensors. *Biotechnol. Bioeng.* 19, 1535, 1977.

CHAPTER 14

Analytical Method for Biomonitoring of Aniline in Urine and Comparison with the Classical Colorimetric Method

Robert Orth, Gary Spies, Rashmii Nair, and Jay M. Wendling

INTRODUCTION

Biomonitoring is a useful tool for evaluating total exposure to a given chemical. Obtaining biomarker data can provide unique opportunities to decide routes of exposure other than the lungs, evaluate the adequacy of personal protective equipment, and appraise the controls in place to individuals from exposure.[1,2] The biological markers used in appraising exposure can range from direct measurement of a health effect to the direct analytical determination of a chemical or its metabolite. The most widely used biomarkers are those that measure the chemical or its metabolite in selected biological specimens such as urine or blood. Even with the measurement of the concentration of the chemical or its metabolite, it is not possible to understand internal dose and exposure relationships without understanding the pharmacokinetics. The points that should be considered in undertaking a valid exposure investigation include the metabolic pathway, the matrix to be sampled, sampling and sample integrity, method of analysis and quantitation, and the use of the results.

To be sure that the aforementioned points can be covered, it is often beneficial to have a team effort in setting up and carrying out a study. There are many key disciplines that should be used, but frequently the team should consist of a toxicologist, medical doctor, industrial hygienist, and analytical chemist. The toxicologist provides the expertise in understanding the possible toxicology and pharmacokinetics of the compound or metabolite to be used as the biomarker. The industrial hygienist is essential for workplace testing since the hygienist is familiar with the exposure possibilities and the on-going monitoring methods such as skin patches or personal and area air sampling. The physician provides the medical expertise and guidance in terms of medical effects and communication. The analytical chemist usually is approached after deciding on a study. The advantage in having a chemist involved from the beginning is in providing guidance as to whether the required detection limits can be obtained. If existing methods will not yield the required information then the analytical chemist can explore what possibilities exist for developing new methods. It is always important to be sure that the information to be gathered will provide enough data to allow conclusions to be drawn. Most of this discussion will be directed toward the analytical aspects of a study that was interested in the exposure to aniline.

Aniline is absorbed by the lungs and the skin and is predominantly metabolized to p-aminophenol (PAP), which is excreted chiefly as conjugates of sulfates and glucuronides.[3-6] This means that if work shift exposure is of interest sampling should occur at the end of the shift. Studies also indicate that in rats less than 1% of the absorbed dose is excreted unchanged in urine.[7] To biomonitor for aniline, urinary levels of PAP are measured. The current biological exposure index (BEI) for aniline is 50 mg of PAP/g creatinine.[8] This BEI was obtained using exposure data of volunteers[7] providing a correlation with TLV based on prevention of methemoglobinemia. The excretion half-life is from 3 to 5 hr, so 89% of PAP is excreted on the day of exposure. Chemicals that may confound aniline biomonitoring include certain household dyes that may contain aniline, certain pharmaceuticals, such as acetanilid,[10] and certain herbicides, such as Propham, Carbetamide, Fenuron, and Siduron, which may be metabolized to PAP. It is also known that 80% of the acetaminophen dose will be conjugated in the liver to yield O-glucuronides and O-sulfates, which upon hydrolysis will yield PAP.[11] The fact that acetaminophen will yield PAP can be beneficial in the development and comparison of analytical methods.

The analytical method plays a key role in the evaluation of exposure. Any new method that is developed should be related to the analytical method that was used in the original adsorption and exposure studies. Also, the understanding of the health effects often sets the limits of detection of a chemical or its metabolite. As the understanding changes, the best detection limits obtainable may be needed. This has occurred for aniline where a recent report of excess number of bladder cancers in workers exposed to o-toluidine and aniline[11] suggested that the exposure to aniline should be determined at the best available detection limits. The analytical methods in that study used high-performance liquid chromatography (HPLC) followed by electrochemical detection of aniline[12] in urine to decide exposure. For the analytical method developed in this report, PAP was the chosen analyte since studies that determined adsorption of aniline used PAP as the biomarker. This is necessary to interpret data accurately for exposure assessment.

A highly specific and sensitive method is needed for the chemical or metabolite that is to be monitored. In the determination of exposure to aniline and the comparison to the BEI, a semiquantitative analytical method[8] that measures PAP in urine by a colorimetric method is presently suggested. Besides the semiquantitative nature of the method, it has other disadvantages, such as poor sensitivity, nonspecificity, and being manual. With the requirement that exposures be as low as possible it is necessary to have detection limits less than 5 µg/mL whereas the current method has detection limits of 5–10 µg/mL of urine. The method is also subject to interference due to the colorimetric approach. An additional error can arise due to systematic errors because of the many manipulations required in the method. To overcome these difficulties, an HPLC method was developed and compared to the colorimetric method.

Sample stability is important and should be checked before a study begins. This is even the case for a well-studied chemical such as PAP. The BEI overview suggests that PAP does not require special handling requirements. Data will be presented which show that PAP was lost from the urine over time.

ANALYTICAL METHODS

Sample Stability

Urine samples that were prepared from freeze-dried urine were spiked with PAP. The samples were stored at 4°C and the amount of PAP was then monitored over a period of up to 2 weeks. The change in concentration was compared to freshly prepared standards. It was proposed that the pH conditions of the urine could be modified to stabilize the changes in the PAP concentration. To test this hypothesis PAP-spiked urine that contained citric acid at a concentration of 0.1 M was also handled in the same manner as the samples containing no citric acid.

Samples

Samples for validation and comparison of colorimetric method with the HPLC method were obtained from individuals who had taken 650 mg of acetaminophen. The first postdose urine sample was collected. Urine samples were also obtained from the same individuals before they were dosed. Samples contained citric acid and were stored at –20°C until analysis.

Hydrolysis of Urine Samples

Common to the two methods discussed here is the hydrolysis of the samples before analysis. To maintain a comparative approach between methods the hydrolysis method described for the colorimetric method was used. The hydrolysis procedure was to take the volume of urine to be used by each method and add concentrated HCl. The colorimetric determination required 10 mL of urine to which 5 mL of HCl was added. The HPLC determination of PAP required 1 mL of urine to which 200 µL of concentrated HCl was added. The hydrolysis was then allowed to proceed at 100°C for 1.5 hr. After the hydrolysis the samples were handled according to the appropriate method.

Determination of *p*-Aminophenol by Colorimetric Analysis

The procedure was the same as that reported by Piotrowski[8] with no changes. The hydrolyzed sample was filtered and diluted with 100 mL of water. To 2 mL of the hydrolyzed dilute urine 5 mL of 5% phenol was added along with 2 mL NH_4OH to form the indophenol blue. The sample was allowed to develop for 30 min at which time the absorbance was read utilizing a Cary UV/VIS spectrophotometer at a wavelength of 630 nm.

Determination of *p*-Aminophenol by HPLC

After completion of the hydrolysis the pH was adjusted to 3–4 with 25% NaOH. This sample was then filtered using a 0.2-μm syringe filter. This filtered sample was loaded into a vial and placed in an HPLC autosampler. The sample was then chromatographed on two Regis ODSII 250 ×4.6 mm C_{18} columns with 5-μm packing.

The mobile phase was 0.4 M KH_2PO_4 at pH 7.5 at a flow rate of 1.0 mL/min. A gradient was used in which the solution was changed with the addition of acetonitrile over 10 min until a 1% acetonitrile concentration was obtained. The run was then held at this concentration for 15 min. The UV detector was set at 301 nm.

Determination of Creatinine

Because the final comparison is normalized to creatinine, a Stanbio Kit method based on a modified Jaffe reaction was used to find the creatinine concentration. Though this is a standard method, a validation was carried out to examine the variation in results.

RESULTS AND DISCUSSION

The stability study results can be seen in Figure 14.1. The percent recoveries for the samples are indicated over a 16-day period. The curve for urine without the citric acid clearly shows that there is substantial loss within the first 5 days. The loss continues through day 16 to give a 36% recovery. With the addition of citric acid the loss is minimized and after day 16 there is still greater than 90% recovery of PAP. This contrasts with the statement made in the BEI guidance in which no special handling is required for PAP in urine.

To validate the new procedure a series of standards made in distilled water were analyzed and compared to a set of urine samples spiked at concentrations equal to the standards. The data obtained from this approach were plotted and a linear regression curve was obtained for each set of data. This comparison provides information through the slope and the intercept of the two curves. If

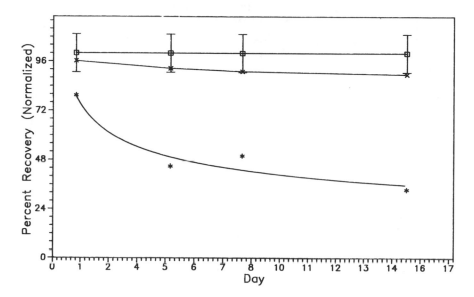

Figure 14.1   Variation of PAP concentration with time.

the slopes of the curves are statistically different it indicates the possibility of a matrix effect. The magnitude of the differences can give guidance as to the cause of the matrix effect. The intercept should reflect the concentration of PAP in the urine before spiking, because the spiking is essentially standard addition. This validation approach was carried out for both methods.

Figure 14.2 shows the results for the spike comparison using the colorimetric method. This method does not have any mechanism for determining recovery and therefore no recovery corrections were applied. The intercepts were the same within the error of the measurements. The value for the slope of the curve for the spiked urine was 4.5×10^{-3} mL absorbance unit per μg whereas the slope for the standards was 5.1×10^{-3} mL absorbance unit per μg. The more positive slope of the urine samples can be easily seen at the higher concentrations of PAP. This would suggest that there is a difference caused by handling or by the matrix. At the lower concentration range of 5 to 20 μg/mL PAP the difference between the two curves is not as large.

Figure 14.3 shows the HPLC chromatogram for an unspiked urine sample and a sample containing 10.7 μg/mL. The total sample run time including column cleanup is less than 1 hr. The detection limit is less than 5 μg/mL. Figure 14.4 shows the results of the spike comparison for the HPLC method. The plot here was peak area vs. concentration of PAP. The slope of the standard

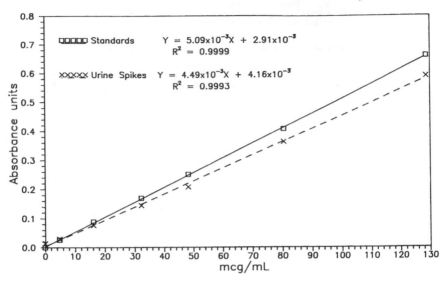

Figure 14.2   Comparison of standard curve to spiked urine samples using the colorimetric method.

curve of 6.76×10^4 µg/mL per peak area did not differ significantly from the standard addition curve obtained for the urine samples of 6.83×10^4 µg/mL per peak area.

The difference between the two methods and the standard curves could arise for many reasons. Probably one of the more important possibilities is the difference in the indol formation as concentrations increase in the colorimetric method. Because the HPLC method has no derivatization and measures the urine by retention time and absorbance, it has a higher degree of specificity. An additional benefit is that the method has fewer manipulations, which lessen the chance for systematic errors.

To adequately test and make comparisons between the two methods, it is necessary to analyze human samples that contained PAP. Because an acetaminophen dose will yield glucuronides and sulfates in urine, which after hydrolysis will yield PAP, it was used in the method validation. The use of acetaminophen provided several advantages. First, it allowed for the determination of the amount of interference that would occur from a normal dose of acetaminophen. It also allowed for a check of the hydrolysis step on a real sample. Finally, the BEI as well as exposure data are based on the colorimetric approach. Therefore, to draw conclusions from any study to determine exposure, a comparison of the two analytical methods is essential.

The samples were collected from individuals who had taken 650 mg of acetaminophen. A predose urine sample was obtained followed by the collection of the first sample after the dose. To analyze the urine from the individuals, it was necessary to dilute the samples. This dilution brought the levels within the

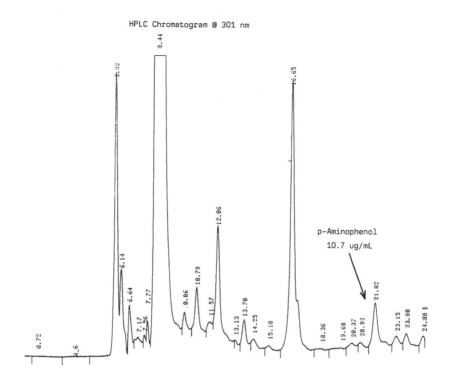

Figure 14.3   HPLC chromatogram of urine containing 10.7 µg/mL of PAP.

lower end of calibration curves of the standards so that the differences between the standard and spiked urine noted previously would be minimized. The results are shown in Table 14.1. Sample M1 predose shows very good comparison between the two methods. The sample labeled S1 predose showed a level of 8.8 mg/g creatinine for the colorimetric and 7.7 mg/g creatinine for the HPLC method. This level is similar to what has been reported[11] as background levels. The M1 predose was repeated to check on reproducibility. The M1 predose is higher than what has been reported as background showing that this individual had an interferant with the analysis or was exposed to a PAP-producing agent. Considering the history of this person, there was no prior indication of exposure to aniline. Notice that after the dose of acetaminophen, the levels for the subjects were greater than 400 mg/g creatinine. The M1 postdose sample had a value of 520 mg/g creatinine whereas the HPLC gave a result of 450 mg/g creatinine. The colorimetric method for the postdose levels always showed a higher level than the HPLC method. Considering the precision of the measurements, this difference is not significant.

The limitation of the colorimetric method and the HPLC method is the background level of approximately 5 µg/mL. The HPLC method has a higher

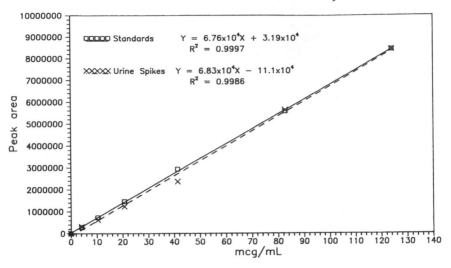

Figure 14.4   Comparison of standard curve to spiked urine samples using the HPLC method.

Table 14.1   **Comparison of Methods from Acetaminophen Dose**

		Results (mg/g creatinine)	
Subject	Sample	Colorimetric	HPLC
M1	Predose	29.2	24.1
M1 (rep)	Predose	29.7	33.2
M2	Postdose	522	452
S1	Predose	8.8	7.7
S2	Postdose	498	490

degree of specificity than the colorimetric method since it combines separation methods with UV detection. However, it is not specific enough to demonstrate that the background is indeed due to PAP. To obtain a higher degree of specificity and even lower detection limits a GC/MS method that uses isotopically labeled PAP should be used. Indeed, the approach to the study that was finally conducted and the results that have been reported[13] used the HPLC method as a screening method followed by a GC/MS method for verification of concentrations above a specified level. This double approach is similar to the approach used in drug screening programs where the less expensive approach, which is not as specific, is used first followed by confirmation by the more highly specific GC/MS method.

In summary, an HPLC method was developed for the biomonitoring of aniline by determining the amount of PAP in urine. The method was found to be comparable to the colorimetric method that was used in determining exposure so that the values obtained can be compared directly with the BEI and TLV values reported in the literature. The method was easy and did not rely on derivatization for the detection of PAP. The limitation appears to be a background level of PAP in urine of around 5 µg/mL. Lastly, the development of the method was decided upon by the interaction of a team approach to a worker-exposure study prior to the study. This approach ensured that the analytical could provide the data required to draw conclusions on exposure.

REFERENCES

1. Lauwerys, R. R. *Industrial Chemical Exposure: Guidelines for Biological Monitoring*. Biomedical Publications, Davis, CA, 1983.
2. Bernard, A. and Lauwerys, R. *Biological Monitoring of Exposure to Chemicals*. Ho, H. M. and Dillon, H. K., Eds. John Wiley, New York, 1987, pp. 1–16.
3. Dutkiewicz, T. Aniline vapors absorption in men. *Med. Pracy.* 12, 139, 1961.
4. Piotrowski, J. Exposure tests for organic compound in industrial toxicology, DHEW (NIOSH) Pub. No. 77-144. U.S. Government Printing Office, Washington, D.C., 1977.
5. Baranowska-Dutkiewicz, B. Skin absorption of aniline from aqueous solutions in man. *Toxicol. Lett.* 10, 367, 1982.
6. Dutkiewicz, T. and Piotrowski, J. Experimental investigations on the quantitative estimation of aniline absorption in man. *Pure Appl. Chem.* 3, 319, 1961.
7. American Conference of Government Industrial Hygienists. *Aniline*, BEI-35, 6th ed., Vol. III. ACGIH, Cincinnati, OH, 1991.
8. Piotrowski, J. Quantitative estimation of aniline absorption through the skin in man. *J. Hyg. Epid. Micro. Immunol.* 1, 23, 1957.
9. Lowery, L. K. The biological exposure index: its use in assessing chemical exposure in the work place. *Toxicology* 47, 55, 1987.
10. Goodman, L. S. and Gilman, A. *The Pharmacological Basis of Therapeutics*, 4th ed. Macmillan, New York, 1970.
11. Ward, E., Carpenter, A., Markowitz, S., Roberts, D., and Halperin, W. Excess number of bladder cancers in workers exposed to ortho-toluidine and aniline. *J. Natl. Cancer Inst.* 83(7), 501, 1991.
12. Settler, L. E., Savage, R. E., Brown, K. K., Cheever, K. L., Weigel, W. W., DeBord, D. G., Teass, A. W., Dankovic, D., and Ward, E. M. Biological monitoring for occupational exposures to ortho-touidine and aniline. *Scand. J. Work Environ. Health* 18 Suppl. 2, 78, 1992.
13. Spies, G., Nair, R. S., Forbes, S. A., Blank, T. L., Orth, R. G., and Wendling, J. M. Use of biological monitoring to evaluate worker exposure to aniline and xylene, American Industrial Hygiene Conference and Exposition, 1992, p. 57.

CHAPTER 15

Biological Monitoring of Workers Exposed to Triazine Herbicides I: Determination of Simazine and Its Metabolites in Urine

J. G. Guillot, A. LeBlanc, O. Samuel, and J. P. Weber

INTRODUCTION

There are very few studies published on the biological monitoring of workers exposed to triazines.[1-4] The main approach was the determination of unchanged triazines in urine but poor correlation was found between the various levels of concentrations of triazines in air and urine.[4] The studies of metabolism of triazines in animals suggested that N-dealkylation of methyl, ethyl, or isopropyl group in the 4 and 6 position was the main route for the elimination of triazines.[4-8] For simazine, the main metabolites would be the 2-chloro-4,6-diamino-s-triazine (metabolite III) and the 2-chloro-4-amino-6-(ethylamino)-s-triazine (metabolite I). In a feasibility study in forestry plant nurseries, we decided to analyze the unchanged simazine and its metabolite III in urine to evaluate the workers' exposure, since metabolite I was not available for the study. Several analytical techniques, including gas chromatography[1,4,7-9] or liquid chromatography,[2,5,6,8-12] have been used for the determination of the triazines in different matrices. All published techniques consist of analyzing the underivatized triazines. Unfortunately at low levels, the triazines have poor chromatographic properties; therefore, the need to make the derivative to improve the chromatographic separation and sensitivity becomes obvious. In this study, we methylated the primary and secondary amino groups with iodomethane and sodium hydride. The derivatives obtained showed excellent chromatographic properties, and the sensitivity was better than 5 µg/L for all compounds studied by GC-MS in the selective ion-monitoring mode.

0-87371-951-4/95/$0.00+$.50
© 1995 by CRC Press, Inc.

EXPERIMENTAL

Instrumentation

A Hewlett-Packard (HP) 5890 gas chromatograph with a cold on column injector and a 5970 mass spectrometer (GC-MS) system was used. The fused silica capillary column was an HP-1 (25 m, 0.2 mm i.d., and 0.33 μm film thickness). The MS was operated with an open split of 1 mL/min in the electronic impact mode at 70 eV. Each day, the instrument was autotuned with perfluorotributylamine from 10 to 600 amu. The data station was an HP-9133 with a 20-megabyte hard disk. For the chromatographic separation, the temperature conditions of the oven were as follows: initial temperature 40°C for 4 min, then increased by 20°C/min to 200°C and held for 1 min, then increased by 5°C/min to 225°C and held 2 min. The temperature setting for cleaning was 20°C/min to 325°C, and the transfer line was kept at 300°C. The carrier gas was purified helium (99.999%) at a pressure of 100 kPa and a flow rate of 0.6 mL/min at room temperature. For the data acquisition, we monitored five groups of ions as follows: 14.6 to 16.1 min, ions 185.7 and 200.9 for metabolite III; 16.2 to 16.9 min, ions 199.8 and 214.8 for metabolite I; 17.0 to 17.8 min, ions 213.9 and 228.7 for simazine; 17.8 to 18.6 min, ion 227.9 for atrazine; and 18.7 to 20.0 min, ion 227.9 for propazine used as internal standard. At the end of the acquisition, a macroprogram was used for the treatment of raw data.

Reagents and Glassware

All solvents and reagents were of analytical grade. Hexane and ethylacetate were of pesticide grade; diethylether was distilled before use. Iodomethane 99%, dimethylformamide 99+% stored under nitrogen, and dry sodium hydride 97% were obtained from Aldrich. Simazine 99%, atrazine 99%, and propazine 99% were purchased from Supelco. The 2-chloro-4,6-diamino-s-triazine 97% (metabolite III) was purchased from Aldrich. All glassware was rinsed with distilled water and methanol and was dried before use.

Urine Collection

Urine samples were collected at the end of workshifts and refrigerated without preservatives at 4°C. Also, the first urine of morning after 3 days without exposure was collected by each worker to evaluate the persistency of simazine in humans. At the arrival at the laboratory, the urines were stored at −80°C until the time of analysis.

Extraction Procedure

Into a 12-mL screw cap glass tube were pipetted 2.0 mL of urine (or standard in normal urine) and 0.1 mL of propazine at 10 mg/L in distilled water. About 25 mg of sodium carbonate was added to adjust the pH to 11 ± 1. The

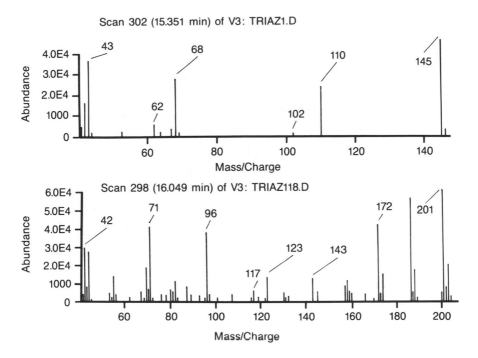

Figure 15.1   Mass spectras of underivated metabolite III of triazine (upper) and its methylated analog (lower).

triazines were extracted from the urine with 10 mL of a mixture of ethylacetate:diethylether:hexane (5:3:2) for 2 min by manual shaking. After centrifugation, 8.0 mL of the organic phase was transferred in a 10-mL Quickfit pear-shaped flask and evaporated to dryness.

For the derivatization, 100 µL of N-dimethylformamide, 200 µL of iodomethane, and about 10 mg of NaH were added to the flask. The mixture was vortexed during 15 sec and left standing at room temperature for 1 hr. The methylated triazines were then dissolved with 10.0 mL hexane:ethylacetate (9:1). The excess sodium hydride was removed by filtration of the organic phase. The excess of iodomethane was destroyed by adding 1 mL of distilled water and about 25 mg of sodium carbonate. After shaking and centrifugation, 5.0 mL of the organic phase was evaporated to dryness. The final extract was reconstituted with 200 µL of hexane:ethylacetate:pyridine (8:2:0.1), and 1 µL of specimen was injected into the GC-MS.

RESULTS

Analytical Method

Mass spectra are shown in Figure 15.1 for underivatized metabolite III and its methylated analog. Mass spectra of the derivatives of simazine, atrazine, and

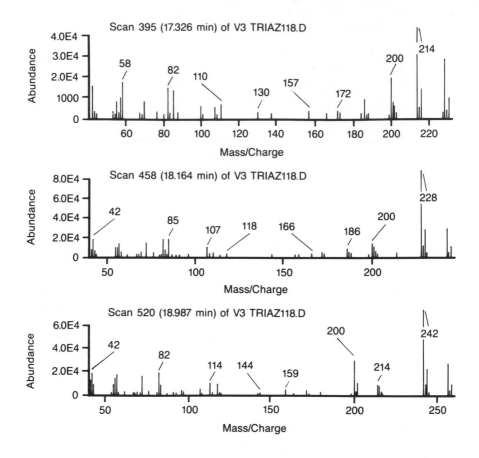

Figure 15.2   Mass spectras of methylated simazine (top), atrazine (middle), and propazine (bottom).

propazine are shown in Figure 15.2. All of the triazines studied showed a distinctive molecular ion, proving that the derivatization was successful. The analysis of the chromatograms acquired in scanning mode (40–300 amu) indicated that the reaction of triazines was complete since underivatized triazines were not found. A chromatogram of a urine extract at 250 µg/L of metabolite III, simazine, and atrazine is presented in Figure 15.3. All peaks are symmetrical, and no adsorption is detected. The detection limits were 5, 2, and 2 µg/L, respectively, for metabolite III, simazine, and atrazine.

PRECISION AND REPRODUCIBILITY

Tables 15.1 and 15.2 show the recovery and the within-day and within-project reproducibility for metabolite III, simazine, and atrazine.

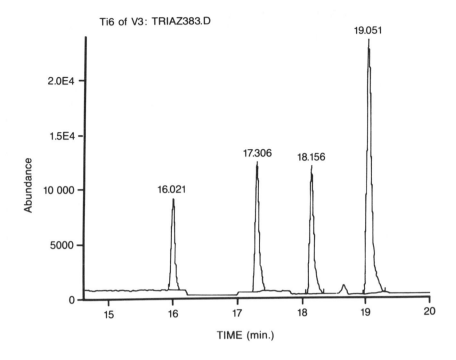

Figure 15.3   Ion chromatograms of urine extracts containing metabolite III, simazine, and atrazine at 250 µg/L each, and propazine at 500 µg/L after derivatization.

Table 15.1 Within-Day Recovery and Precision[a]

Triazines	Concentration (µg/L)		
	Added	Found	CV%
Simazine	25	26.1	10.0
Simazine	50	50.2	7.6
Atrazine	25	27.7	11.9
Atrazine	50	50.2	6.8
Metabolite III	25	22.8	7.0
Metabolite III	50	56.9	7.9

[a] Based on five determinations of triazines added to normal urine.

Table 15.2 Within-Project Recovery and Precision[a]

Triazines	Concentration (µg/L)		
	Added	Found	CV%
Simazine	50	47.0	15
Atrazine	50	52.9	15
Metabolite III	50	48.9	21

[a] Based on 23 determinations of triazines added to normal or workers urine during 45 days of analysis.

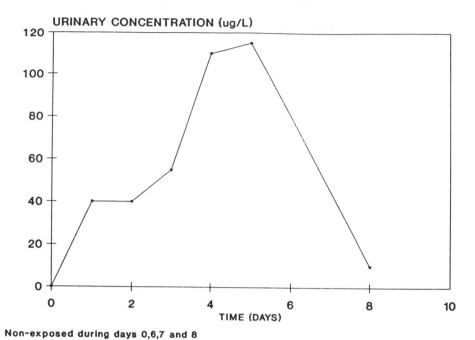

Non-exposed during days 0,6,7 and 8

Figure 15.4   Urinary concentration of metabolite III vs. the time for a worker exposed to simazine.

PLANT NURSERY STUDY

During the summers of 1989 and 1990, two studies were carried out in order to evaluate professional exposure of workers using triazine herbicides in nurseries. Overall, 163 urine samples were analyzed, with 131 samples collected after each work day and 32 samples collected on mornings following 3 days of nonexposure.

As reported in previous studies,[1,4] the urinary levels of unchanged simazine were low. Simazine was detected only in 4.6% of all the urine samples and, in some cases, reached a concentration of no more than 5 µg/L. We found that the urinary concentration of simazine was not correlated with the type of work or the protective gear used. We therefore concluded that this measurement would not be suitable for the evaluation of exposure to this triazine.

The determination of the urinary levels of the metabolite III is much more interesting for the evaluation of worker exposure. Figure 15.4 shows a typical case of the urinary levels of metabolite III vs. time. The urinary levels of metabolite III are not detectable before exposure, but they increase for each work day and are generally completely eliminated after 3 days without exposure. Moreover, the urinary levels are correlated with the various workstations (Fig. 15.5). It is also seen that the type of application influences the urinary level. The detailed analysis of these results will be published elsewhere.

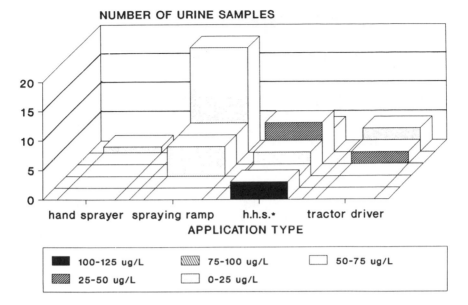

NUMBER OF URINE SAMPLES

| 100-125 ug/L | 75-100 ug/L | 50-75 ug/L |
| 25-50 ug/L | 0-25 ug/L | |

* hydrolic hand sprayer behind
the tractor

Figure 15.5   Urinary levels of metabolite III vs. pulverization technique: hand sprayer, spraying
ramp, hydrolic hand sprayer behind the tractor, vehicle driver (hand sprayer).

For the interpretation of the urinary concentrations of metabolite III found
in the urine of workers, two volunteers ingested 0.5 and 1 mg of simazine
corresponding to 3.5 and 7 times the acceptable daily intake.

This study showed that simazine was undetectable in the urine of the two
volunteers; however, the urinary concentrations of metabolite III (Fig. 15.6)
indicated that this metabolite would be a good biological indicator of exposure.
For the volunteer who absorbed 0.5 mg, 14.8% of simazine was metabolized to
metabolite III and for the volunteer who absorbed 1 mg, 8.8% was metabolized.
The half-life of metabolite III calculated from the amounts excreted by a one
compartment model was found to be 12 hr.[13]

DISCUSSION

Metabolite III of the triazines is very difficult to quantify in trace analysis.
It is a relatively polar compound, insoluble (<10 mg/L) in hexane, diethyl ether,
dichloromethane, and chloroform. It is very slightly soluble in water and 0.1 M
HC1 ($\approx$50 mg/L) and also tetrahydrofuran ($\approx$100 mg/L). It is partially soluble in
methanol ($\approx$200 mg/L), dimethylformamide, or dioxane ($\approx$500 mg/L). The mixture of organic solvents used for the extraction was optimized to extract the

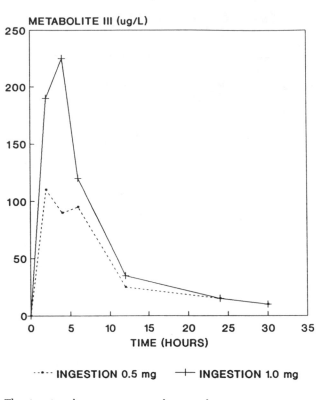

Figure 15.6   The simazine elimination curve in the two volunteers.

triazines and their metabolites at levels inferior to 1 mg/L. Recoveries exceed
90% in this range of concentration.

The derivatization of the species (triazines and metabolites) is difficult to
accomplish. Acetylation of metabolite III by acetic anhydride, trifluoroacetic
anhydride, or the N-methyl-bistrifluoroacetamide (MBTFA) is not favored even
in the presence of a catalyst and at high temperatures. Acetylation being a
general and most common reaction to derivatize amines, we can explain this
unusual behavior from the resonance structures between the aromatic ring and
the amino groups, as proposed by Pearlman and Banks.[14]

The formation of a trimethylsilyl derivative was successfully accomplished
for metabolite III. This type of derivatization was rejected since silanization of
secondary amines is not possible for metabolite I, simazine, atrazine, and
propazine resulting in chromatographic problems at low levels.

Methylation of aromatic amines with iodomethane is seldom used since
reaction rates are slow, not reproducible, and cannot be performed at room
temperature. The addition of sodium hydride reduces amino groups, creating
anions of the type RNH^- or $R_1R_2N^-$. This anion can then be substituted with the
iodide of iodomethane. With quantities of 1 μg or more of the triazines or their
metabolites, the reaction is spontaneous at room temperature. The first advan-

tage of this procedure is that all species have the same polarity after derivatization, and their solubilities in organic solvents are increased. The second advantage is found during the chromatography where the methyl analogs will elute at about the same temperature. Furthermore, methylation will increase the molecular weight of the compound, which is interesting in the selected ion monitoring mode. The methylated urine extract resolubilized in organic solvents prior to injection are clear and colorless and can be used without other means of purification. After many injections ($\approx$200), no significant increase in the lens potential of the ion focus was observed, indicating that the extracts do not contaminate the elution source of the GC-MS. Furthermore, the properties of the GC column did not vary during the course of the analyses. Possible chromatographic interferences from the urine matrix were evaluated by the use of pooled urine samples of smokers and coffee drinkers. No interferences were observed during these analyses. Mass spectrometry in the selected ion mode provided the necessary sensitivity and selectivity to quantify the species at very low levels of concentration.

CONCLUSION

We have developed a new method, reliable and sensitive, that allows detection of simazine and its metabolite III in urine. The derivatization procedure is adequate for amines (except tertiary), hydroxy, and carboxylic acid functional groups. Judging from the results obtained from the workers and the two volunteers, we can conclude that the concentrations of metabolite III are excellent indicators of exposure to simazine compared to poor correlations found with the simazine concentrations.

ACKNOWLEDGMENTS

The authors are grateful to Dr. D.R. Ridley from Ciba-Geigy, and Denise Phaneuf, Francine Labrecque, and Denise Langlois of Centre de Toxicologie du Quebec for their professional or technical assistance during this study completed in collaboration with the Ministry of Forests of Quebec.

REFERENCES

1. Catenacci, G., Maroni, M., Cottica, D., and Pozzoli, L. Assessment of human exposure to atrazine through the determination of free atrazine in urine. *Bull. Environ. Contam. Toxicol.* 44, 1, 1990.
2. Pommery, J., Mathieu, M., Mathieu, D., and Lhermitte, M. High-performance liquid chromatographic determination of atrazine in human plasma. *J. Chromatogr.* 526, 569, 1990.

3. Reed, J. P., Hall, F. R., and Krueger, H. R. Measurement of ATV applicator exposure to atrazine using an ELISA method. *Bull. Environ. Contam. Toxicol.* 44, 8,1990.
4. Ikonen, R., Kangas, J., and Savolainen, H. Urinary atrazine metabolites as indicators for rat and human exposure to atrazine. *Toxicol. Lett.* 44, 109, 1988.
5. Bakke, J. E., Larson, J. D., and Price, C. E. Metabolism of atrazine and hydroxyatrazine by the rat. *J. Agric. Food Chem.* 20, 602, 1972.
6. Larsen, G. L. and Bakke, J. E. Metabolism of 2-chloro-4-cyclopropylamino-6-isopropylamino-s-triazine (cyprazine) in the rat. *J. Agric. Food Chem.* 23, 388, 1975.
7. Bradway, D. E. and Moseman, R. F. Determination of urinary residue levels of N-dealkyl metabolites of triazine herbicides. *J. Agric. Food Chem.* 30, 244, 1982.
8. Erickson, M. D., Frank, C. W., and Morgan, D. P. Determination of s-triazine herbicide residues in urine: Studies of excretion and metabolism in swine as a model to human metabolism. *J. Agric. Food Chem.* 27, 743, 1979.
9. Rostad, C. E., Pereira, W. E., and Leiker, T. J. Determination of herbicides and their degradation products in surface waters by gas chromatography/positive chemical ionisation/tandem mass spectrometry. *Biomed. Environ. Mass Spectrom.* 18, 820, 1989.
10. Frohlich, D. and Meier, W. HPLC determination of triazines in water samples in the ppt-range by on-column trace enrichment. *J. High Resol. Chromatogr.* 12, 340, 1989.
11. Ferris, I. G. and Haigh, B. M. A rapid and sensitive HPLC procedure for the determination of atrazine residues in soil-water extracts. *J. Chromatogr. Sci.* 25, 170, 1987.
12. Battista, M., Corcia, A. D., and Marchetti, M. Extraction and isolation of triazine herbicides from water and vegetables by a double trap tandem system. *Anal. Chem.* 61, 935, 1989.
13. Gibaldi, M. and Perrier, D. *Pharmacokinetics.* Marcel Dekker, New York, 1982, p. 40.
14. Pearlman, W. M. and Banks, C. K. Substituted chlorodiamino-s-triazines. *J. Am. Chem. Soc.* 70, 3726, 1948.

CHAPTER 16

Biological Monitoring of Workers Exposed to Triazine Herbicides II: Presentation of a Method for the Determination of Hexazinone and Its Main Metabolites in Urine of Exposed Forestry Workers

A. LeBlanc, O. Samuel, J. G. Guillot, and J. P. Weber

INTRODUCTION

Hexazinone [3-cyclohexyl-6-(dimethylamino)-1-methyl-1,3,5-triazine-2,4-(1H,3H)-dione] is commercially available from DuPont® under the name of VELPAR weed killer. In the province of Quebec, the Ministry of Forests uses VELPAR to control weed growth prior to replanting operations.

When absorbed in the body, hexazinone is relatively quickly eliminated in urine and, for the most part, as metabolites A, B, and C[1-2] (Fig. 16.1). The major metabolite (B) indicates that dealkylation of hexazinone is the major elimination pathway in humans followed by dealkylation and hydroxylation (C) and finally hydroxylation alone (A). We have developed a method to quantify the amount of hexazinone and its three main metabolites possibly present in the urine of exposed forestry workers. We hereby propose an analytical method for the determination of these species.

ANALYTICAL METHOD

Extraction Procedure

The determination of hexazinone and its three main metabolites is carried out using a 2 mL urine sample. After adding propazine (1 µg) as an internal standard, the pH is adjusted to 8.5 with sodium bicarbonate. To improve

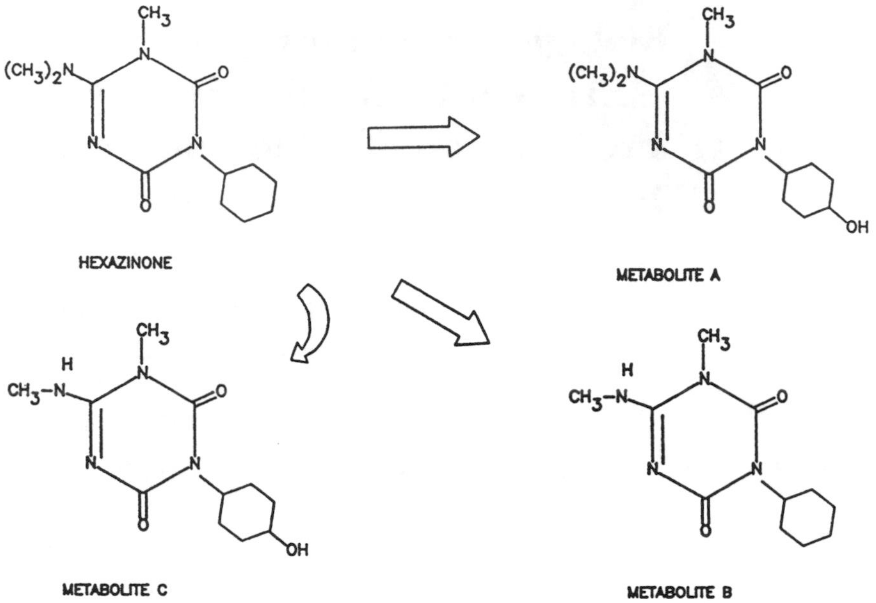

Figure 16.1   Metabolism of hexazinone.

recoveries, the urine sample is saturated with sodium chloride. The actual extraction is achieved by a 4-fold extraction with chloroform. After removal of the organic phase, a fraction is evaporated to dryness. At this stage, another evaporation step is included to completely remove residual chloroform vapors that would interfere in the derivatization that follows. We add 200 μL of ethyl acetate and evaporate to dryness. To the dry residue, 100 μL of dimethyl-formamide is added to solubilize the species prior to derivatization.

Derivatization Procedure

Using sodium hydride (40 mg), reduction of the primary and secondary amino groups and hydroxyl functional groups is accomplished [Eq. (1)]. From this reaction, the corresponding sodium salts are produced, releasing hydrogen. After adding an excess of iodoethane (200 μL), the corresponding ethyl derivatives are formed and sodium iodide remains in solution. The reaction time is 1 hr at room temperature.

$$
\begin{bmatrix}
\begin{pmatrix} R-O-H \\ R_1R_2-N-H \end{pmatrix} + NaH \rightarrow \begin{pmatrix} R-O^--Na^+ \\ R_1R_2-N^--Na^+ \end{pmatrix} + H_2 \\[2em]
\begin{pmatrix} R-O^--Na^+ \\ R_1R_2-N^--Na^+ \end{pmatrix} + C_2H_5I \rightarrow \begin{pmatrix} R-O\ -C_2H_5 \\ R_1R_2-N-C_2H_5 \end{pmatrix} + NaI
\end{bmatrix}
\tag{1}
$$

Following the derivatization procedure, reaction by-products and excess chemicals (NaH and CH_3CH_2I) are removed to prevent premature deterioration of the GC column. To the reaction mixture, 10 mL of hexane:ethyl acetate (9:1) is added and agitated on a vortex mixer for a few seconds. The organic solvents are filtered on top of 1 mL of a 5% solution of sodium carbonate in water. After vigorous shaking for 2 min followed by centrifugation, a fraction of the organic phase is evaporated to dryness. The residue is taken in 200 µL of isooctane:pyridine (10:0.1) ready to inject in the GC-MS.

HYDROLYSIS

To test if hydroxy metabolites (A and C) were excreted as glucuronic acid conjugates, we carried out several types of hydrolysis on urine samples obtained from the most exposed worker. We performed acid hydrolysis using hydrochloric acid at different temperatures and for different periods of time. Also, using β-glucuronidase pastella vulgata type L-II, we performed a 24-hr enzyme hydrolysis (Table 16.1).

INSTRUMENT PARAMETERS

Mass detection is accomplished using a Hewlett Packard quadrupole mass spectrometer model 5970 coupled to a gas chromatograph model 5890 equipped with a cold on column injector. A 25-m, 0.2-mm i.d. HP-1 column is used (film thickness 0.3 µm). The carrier gas is helium and is maintained at a pressure of 100 kPa in the column.

The temperature program (Table 16.2) and identification parameters (Table 16.3) are given.

METHOD EVALUATION

During each run, five standards are analyzed in sequence to construct a calibration curve. A blank urine sample is spiked with hexazinone and metabolites A, B, and C to give the following concentrations: 0, 0.1, 0.5, 1.0, and 2.0 mg/L (Fig. 16.2). To control the efficiency of the derivatization procedure, 0.5 µg of each

Table 16.1 Metabolite C

	Normalized response factors			
	HCl hydrolysis		Enzyme hydrolysis	
	With	Without	With	Without
Sample 1	0.89	1	0.80	1
Sample 2	0.96	1	1.09	1

Table 16.2 Temperature Program[a]

Initial temp	Initial time	Rate (°C/min)	Final temp	Final time	Total time
100	4	20	200	1	10
		5	285	0	27
		2.5	305	0	35
		20	325	1	37

[a] Transfer line = 300°C.

Table 16.3 Identification Parameters

	Ion monitored	RT	RRT
Hexazinone	170.9	24.07	1.406
Met A	170.9	28.56	1.668
Met B	184.9	24.90	1.454
Met C	184.9	29.38	1.716
Propazine	255.9	17.12	1.000

Table 16.4 Recovery from a Urine Matrix (range 0–5 mg/L)

	Hexazinone	Met A	Met B	Met C
Slope	0.897	0.925	0.935	1.103
Intercept	0.116	0.144	0.080	0.060
r	0.995	0.988	0.999	0.998

Table 16.5 Detection Limits

	µg/L
Hexazinone	2
Met A	10
Met B	2
Met C	2

species is derivatized. Also, a urine sample from an exposed worker serves as a control to verify the reproducibility of the method. Table 16.4 shows the recovery study and Table 16.5 the detection limits. The recovery was studied between 0.05 and 5 mg/L. A graph of theoretical concentrations vs. obtained concentrations (using the 0.5 mg/L as the point calibrator) is plotted, and the regression analysis details are given. Good recovery is obtained via the slope of the line. Positive intercepts may suggest negligible absorption on the GC column. Linearity is clearly shown by the coefficient of correlation. Detection limits are very low because the species are derivatized and rendered nonpolar.

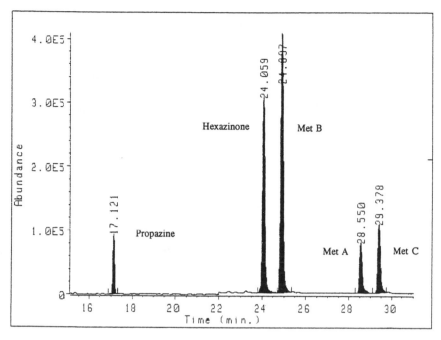

STANDARD (2 MG/L)

Figure 16.2 Spiked urine sample.

Table 16.6 Reproducibility

| | | Concentration (mg/L) | | |
	N	Target	Found	%RSD
Within-day				
Hexazinone	4	1	1.020	4.5
Hexazinone	4	5	4.3	6.0
Met A	4	1	1.045	10.2
Met A	4	5	4.6	16.0
Met B	4	1	0.980	6.9
Met B	4	5	4.7	3.0
Met C	4	1	1.060	15.6
Met C	4	5	5.6	13.0
Long-term				
Hexazinone	15	1.0	1.02	14
Met A	15	1.0	1.12	20
Met B	15	1.0	1.01	12
Met C	15	1.0	1.20	23

Metabolite A has a higher detection limit because its SIM ion is less intense than the SIM ions of the other species. Within-day and long-term reproducibilities are given in Table 16.6.

Percent relative standard deviations do not exceed 23%, which is very acceptable in the context of biological monitoring.

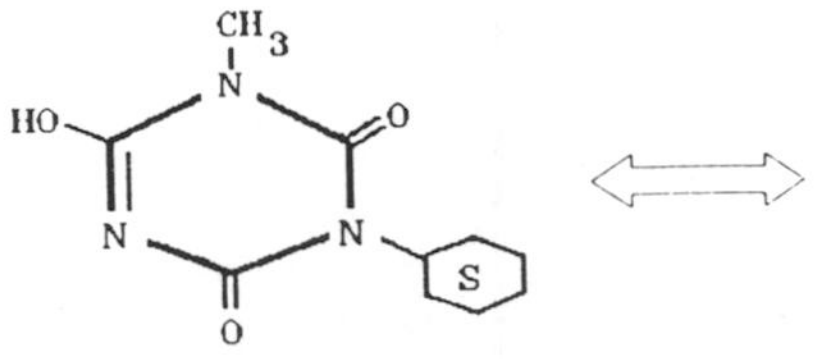

METABOLITE D

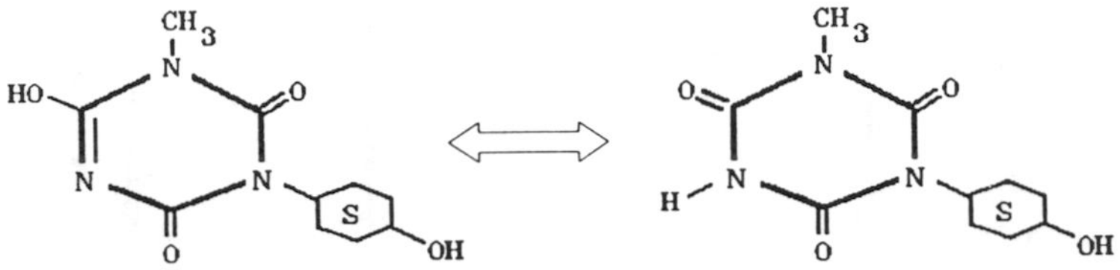

METABOLITE E

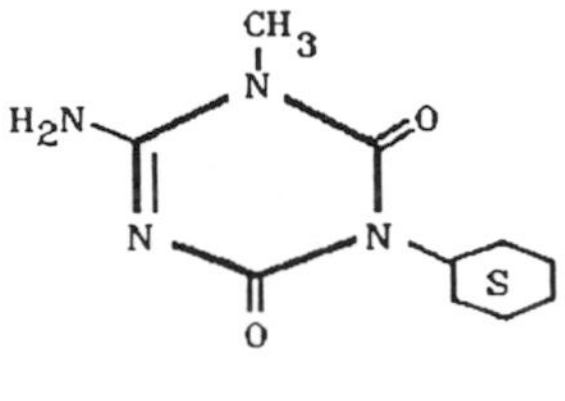

METABOLITE F

Figure 16.3   Minor metabolites.

RESULTS AND DISCUSSION

Since dealkylation (in this case demethylation) seems to be the main route of excretion and metabolite B (the major metabolite) still has one residing methyl group, it would be logical to propose that a second demethylation could take place. If that is the case, another major metabolite could be formed, metabolite F. It was not possible, in this study, to confirm the presence of metabolite F due to the nonavailability of this compound. Attempts have been made to incorporate two other metabolites, although minor, in the analytical method. Metabolite D (Fig. 16.3) has the dominant secondary amine resonance structure and, therefore, responds well in producing only one derivative when ethylated with iodoethane. Metabolite E (Fig. 16.3), however, generated three derivatives with iodoethane, most probably because it had degraded by the time we tested it.

Judging from the hydrolysis test (Table 16.1), metabolite C is eliminated as a nonconjugate (we assume the same for metabolite A). Of course, this cannot be proven because our tests were not performed on an actual conjugated standard compound. One interesting fact that helps to indicate that metabolites A and C are excreted as nonconjugates is that they are very water soluble and conjugation would not really help in increasing their solubility.

CONCLUSION

We have developed a new method, reliable and sensitive, capable of detecting low µg/L levels of hexazinone and metabolites in urine. This method is well suited for biological monitoring with recovery and reproducibility studies well inside acceptable criteria. The method showed that metabolite B in the urine of exposed workers is the best biological indicator of exposure.

REFERENCES

1. Rhodes, R. C. and Jewell, R. A. Metabolism of C-labeled hexazinone in the rat. *J. Agric. Food Chem.* 28, 303, 1980.
2. Reiser, R. W., Belasco, I. J., and Rhodes, R. C. Identification of metabolites of hexazinone by mass spectrometry. *BioMed. Mass. Spec.* 10(11), 581, 1983.

Index